S0-DRI-078

Ernst Schering Research Foundation Workshop 30
Therapeutic Vaccination Strategies

Springer
Berlin
Heidelberg
New York
Barcelona
Hong Kong
London
Milan
Paris
Singapore
Tokyo

Ernst Schering Research Foundation
Workshop 30

Therapeutic Vaccination Strategies

P. Walden, W. Sterry, H. Hennekes
Editors

With 37 Figures and 9 Tables

Series Editors: G. Stock and M. Lessl

ISSN 0947-6075
ISBN 3-540-67298-2 Springer-Verlag Berlin Heidelberg New York

CIP data applied for

Die Deutsche Bibliothek – CIP-Einheitsaufnahme
Schering-Forschungsgesellschaft <Berlin>: Ernst Schering Research Foundation Workshop. - Berlin; Heidelberg; New York; Barcelona; Budapest; Hong Kong; London; Milan; Paris; Santa Clara; Singapore; Tokyo: Springer.
ISSN 0947-6075
30. Therapeutic Vaccination Strategies. - 2000
Therapeutic Vaccination Strategies ; with tables / P. Walden, W. Sterry, H. Hennekes, ed. - Berlin; Heidelberg; New York; Barcelona; Budapest; Hong Kong; London; Milan; Paris; Singapore; Tokyo: Springer, 2000
(Ernst Schering Research Foundation Workshop; 30)
ISBN 3-540-67298-2

Typesetting: Data conversion by Springer-Verlag
Printing: Druckhaus Beltz, Hemsbach. Binding: J. Schäffer GmbH & Co. KG, Grünstadt
SPIN:10751310 21/3134/AG-5 4 3 2 1 0 – Printed on acid-free paper

Preface

The induction of immune responses against tumor cells by vaccination is rapidly evolving as a therapeutic modality with new potentials for the treatment of cancer. It is based on the fact that our immune system can identify tumor cells and, once activated, is capable of developing specific immunity against the neoplastic cells. Numerous observations and intense research clearly document the major contribution of the immune system to the prevention of cancer. And there are many reports of patients suffering from malignant melanoma or other tumors who mount a spontaneous immune response against their tumor cells that results in tumor regression.

Based on the recent advances in our understanding of the components of our immune system, their interactions and the regulation of immune responses, we are now able to design vaccination strategies that induce or enhance cell-mediated immunity against tumors. A major advancement came with the identification and characterization of relevant tumor antigens, which are suitable target structures for anti-tumor immune response. First clinical trials using such vaccine strategies have yielded encouraging results in patients. However, in spite of many reported cases of successful therapy of cancer by vaccination many patients still do not experience relief after such treatments. These initial clinical trials and the accompanying investigations have revealed a number of important results that indicate the direction of future research and development in the field.

With a two-day workshop held in May 1999 in Berlin, Germany, we attempted to join experts from different fields related to therapeutic vaccination in order to evaluate the current status and to discuss the fu-

The participants of the workshop

ture directions of the development of strategies for cancer immune therapy. The topics explored included the basic principles of anti-cancer immunity and of successful vaccination therapy, such as the identification of suitable tumor antigens, adjuvants, or vectors. Recurring themes were the need for and the utilization of immune monitoring techniques which are required for the evaluation of the vaccination effects and as basis for a systematic development of new therapeutic approaches, as well as the clinical criteria for the selection of cancers that are suitable for immune therapy and the clinical settings required. The workshop was organized in four sessions covering the basic concepts, the available technologies, the clinical conditions for and experiences with therapeutic vaccination trials, and the legal considerations for their clinical application. Extending the scope of the workshop, new concepts for the use of vaccination strategies for the elimination or attenuation of autoreactive T cells in autoimmune diseases were discussed as well.

We are indebted to the Ernst Schering Research Foundation for providing us with the necessary resources for this workshop and the support of the representatives of the Foundation in the organization of this meeting. We wish to thank the speakers and the participants for their most valuable presentations and lively discussions at the workshop and the authors of the chapters in this proceedings volume for their contributions. We also thank the chairpersons, Drs. Dominik Mumberg, Ruth Thieroff-Ekerdt, and Michael Töpert of Schering AG, Berlin, Germany, and Dr. David Henderson of Berlex Biosciences, Richmond, Calif., USA, for guiding the sessions.

Peter Walden, Wolfram Sterry, Hartwig Hennekes

Table of Contents

List of Editors and Contributors

Editors

H. Hennekes
Schering AG, Experimental Dermatology, Müllerstrasse 178, 13342 Berlin, Germany

W. Sterry
Department of Dermatology and Allergy, Medical Faculty Charité, Humboldt University, Schumannstrasse 20/21, 10117 Berlin, Germany

P. Walden
Department of Dermatology and Allergy, Medical Faculty Charité, Humboldt University, Schumannstrasse 20/21, 10117 Berlin, Germany

Contributors

S. Brocke
Hebrew University Jerusalem, Hadassah Medical School, Department of Pathology, P.O. Box 12272, 91120 Jerusalem, Israel

J.-C. Cerottini
Division of Clinical Onco-Immunology, Ludwig Institute for Cancer Research Lausanne Branch, University of Lausanne, Lausanne, Switzerland

V. Cerundolo
Molecular Immunology Group, Nuffield Department of Medicine, John Radcliffe Hospital, Oxford, UK

K. Cichutek
Paul-Ehrlich-Institut, Bundesamt für Sera und Impfstoffe,
Abteilung Medizinische Biotechnologie, Paul-Ehrlich-Strasse 51–59,
63225 Langen, Germany

M. Collins
Department of Immunology, Windeyer Institute of Medical Sciences,
46 Cleveland Street, London W1P 6DB, UK

L. Cooper
Fred Hutchinson Cancer Research Center, Division of Clinical Research,
1100 Fairview Avenue N, Seattle, WA 98109–1024, USA

R. de Fries
Fred Hutchinson Cancer Research Center, Division of Clinical Research,
1100 Fairview Avenue N, Seattle, WA 98109–1024, USA

P.D. Greenberg
Fred Hutchinson Cancer Research Center, Division of Clinical Research,
1100 Fairview Avenue N, Seattle, WA 98109–1024, USA

T.J. Hamblin
Molecular Immunology Group, Tenovus Laboratory, Southampton University
Hospital, Tremona Road, Southampton SO9 4XY, UK

G. Herberth
Department of Dermatology and Allergy, Medical Faculty Charité,
Humboldt University, Schumannstrasse 20/21, 10117 Berlin, Germany

J.W. Hodge
National Cancer Institute, Tumor Immunology and Biology Branch, National
Institute of Health, Building 10, Room 5B38, Bethesda, MD 20879, USA

C.A. King
Molecular Immunology Group, Tenovus Laboratory, Southampton University
Hospital, Tremona Road, Southampton SO9 4XY, UK

A.M. Krieg
University of Iowa, Department of Internal Medicine, 540 EMRB, Iowa City,
IA 52242, USA

F. Lejeune
Multidisciplinary Oncology Center, University Hospital, Lausanne, Switzerland

D. Lewinsohn
Fred Hutchinson Cancer Research Center, Division of Clinical Research, 1100 Fairview Avenue N, Seattle, WA 98109–1024, USA

D. Liénard
Division of Clinical Onco-Immunology, Ludwig Institute for Cancer Research Lausanne Branch, University of Lausanne, Lausanne, Switzerland

H. Mutimer
Fred Hutchinson Cancer Research Center, Division of Clinical Research, 1100 Fairview Avenue N, Seattle, WA 98109–1024, USA

M.J. Pittet
Division of Clinical Onco-Immunology, Ludwig Institute for Cancer Research Lausanne Branch, University of Lausanne, Lausanne, Switzerland

J. Rice
Molecular Immunology Group, Tenovus Laboratory, Southampton University Hospital, Tremona Road, Southampton SO9 4XY, UK

S.R. Riddell
Fred Hutchinson Cancer Research Center, Division of Clinical Research, 1100 Fairview Avenue N, Seattle, WA 98109–1024, USA

P. Romero
Division of Clinical Onco-Immunology, Ludwig Institute for Cancer Research Lausanne Branch, University of Lausanne, Lausanne, Switzerland

S.S. Sahota
Molecular Immunology Group, Tenovus Laboratory, Southampton University Hospital, Tremona Road, Southampton SO9 4XY, UK

J. Schlom
National Cancer Institute, Tumor Immunology and Biology Branch, National Institute of Health, Building 10, Room 5B38, Bethesda, MD 20879, USA

D.E. Speiser
Division of Clinical Onco-Immunology, Ludwig Institute for Cancer Research Lausanne Branch, University of Lausanne, Lausanne, Switzerland

M.B. Spellerberg
Molecular Immunology Group, Tenovus Laboratory, Southampton University Hospital, Tremona Road, Southampton SO9 4XY, UK

W. Sterry
Department of Dermatology and Allergy, Medical Faculty Charité, Humboldt University, Schumannstrasse 20/21, 10117 Berlin, Germany

F. K. Stevenson
Molecular Immunology Group, Tenovus Laboratory, Southampton University Hospital, Tremona Road, Southampton SO9 4XY, UK

G. Stingl
Division of Immunology, Allergy and Infectious Diseases, Department of Dermatology, Währinger Gürtel 18–20, 1090 Vienna, Austria

A.R. Thompsett
Molecular Immunology Group, Tenovus Laboratory, Southampton University Hospital, Tremona Road, Southampton SO9 4XY, UK

M. Topp
Fred Hutchinson Cancer Research Center, Division of Clinical Research, 1100 Fairview Avenue N, Seattle, WA 98109–1024, USA

U. Trefzer
Department of Dermatology and Allergy, Medical Faculty Charité, Humboldt University, Schumannstr. 20/21, 10117 Berlin, Germany

D. Valmori
Division of Clinical Onco-Immunology, Ludwig Institute for Cancer Research Lausanne Branch, University of Lausanne, Lausanne, Switzerland

P. Walden
Department of Dermatology and Allergy, Medical Faculty Charité, Humboldt University, Schumannstrasse 20/21, 10117 Berlin, Germany

E.H. Warren
Fred Hutchinson Cancer Research Center, Division of Clinical Research,
1100 Fairview Avenue N, Seattle, WA 98109–1024, USA

D. Zhu
Molecular Immunology Group, Tenovus Laboratory, Southampton University
Hospital, Tremona Road, Southampton SO9 4XY, UK

1 Vaccination Therapy for Malignant Disease: The Clinical Perspective

W. Sterry

1.1 Introduction

We are in the middle of an explosion. Vaccination for the treatment of malignant disease is used exponentially. Numerous academic and non-academic centers have started such protocols, and a wide variety of malignant diseases are studied. The number of variables is high. Variations include type and stage of disease, type of tumor antigen (complete cells, cell lysates, complete natural or recombinant molecules, T cell peptides, naked DNA or transfected DNA), type of antigen presentation system, dose, optimal cytokines and costimulatory molecules, route and frequency of administration, as well as combination with other treatments. The different possibilities to combine each of these variations give astronomic numbers, and it is likely that all these combinations are going to be tested. The question is obvious: should the scientific community try to define some parameters that can be used when planning studies and when selecting and monitoring patients, and how can existing information and guidelines be implemented into the evaluation process of study protocols. In order to make studies comparable, and to exchange data continuously between groups, similar investigations should be performed using identical techniques. Meetings like the *Workshop on Vaccination Strategies* sponsored by the Ernst Schering Foundation held from 27–29 May 1999, in Berlin provide an ideal platform for discussing such issues along with the many scientific aspects relevant for tumor vaccination. The author does not try to give his personal view of the questions raised below. It remains for the reader to answer the questions addressed, or to decide which of these problems can be solved with our current knowledge and which will have to wait for the years to come.

1.2 Clinical Aspects of Tumor Vaccination

Question 1: Selection of Malignant Diseases as Candidate Disease for Vaccination and Patients

Which malignant diseases are most suitable for vaccination therapy, and why. For example, malignant melanoma is thought to be an excellent candidate by many oncological dermatologists, since it is known to be immunogenic by the occurrence of spontaneous regression and simultaneous development of vitiligo (a depigmentation of the skin thought to be due to immune responses against melanocytes). Furthermore, it is possible to monitor and investigate metastases quite easily since metastases develop regularly in the skin. If this is true, what then with renal cell carcinoma or even glioblastoma? Are they characterized by spontaneous immune-mediated regressions? Can metastases be obtained easily? The answer is no, but nevertheless vaccination studies are performed in these and many comparable malignancies. Therefore, the question arises concerning criteria for selection of malignancies that should be tested for tumor vaccination response.

Is it possible to define valid criteria that would allow to estimate the probability for a given malignancy to respond to tumor vaccination? If not, what data would be need to make such estimates, and how could such data be generated. Or would it be wise not to make such recommendations and instead let each type of malignancy be tested by chance?

Question 2: Selection of Patients into Tumor Vaccination Trials

Similar to the problem of identifying malignancies characterized by a high probability to respond to tumor vaccination, is the question concerning the disease stages that should be targeted by tumor vaccine therapy. Currently, most new protocols are tested in patients with metastatic disease, after standard treatments have failed. Quite clearly, everybody in the field knows that metastatic disease with a high tumor burden is not the appropriate setting for immune therapy, since too many tumor cells as well as downregulatory cytokines most likely will prevent the generation of an effective anti-tumor immune response. This obvious contradiction raises the question, which stages or substages are best suited for tumor vaccination. Which are the criteria for selecting such stages or substages. Furthermore, is metastatic disease, even in selected patients with low tumor burden and with active cellular immunity, the appropriate stage for demonstration of "proof of principle" of any new strategy? Clear and widely accepted answers will not be readily at hand, but ways should be defined to come to such answers some day.

Question 3: How Can the Problem of Tumor Heterogeneity be Integrated into Study Designs

Standardized protocols for tumor vaccination will encounter the problem of tumor heterogeneity. This heterogeneity is of relevance both inter- as well as intraindividually (Table 1). Different subtypes (i.e.,

Table 1. The problem of tumor heterogeneity

1. Interindividual heterogeneity
Subtypes of primary malignancy with different biological aggressivity may exist (i.e., nodular versus lentigo maligna melanoma)
Time course of tumor growth or metastatic spread is highly variable
Pattern of metastasis will influence the amount of the immune response that can be generated due to differences in microarchitecture of different organs
2. Intraindividual heterogeneity
Numerous subclones develop during metastatic disease
Can changes in tumor characteristics be integrated into ongoing vaccination?

lentigo maligna melanoma, nodular melanoma, superficial spreading melanoma, acrolentiginous melanoma) of each tumor with different growth and metastatic potential exist, which most likely will respond differentially to treatment. Also, the metastatic preferences differ profoundly between patients, and most likely the anatomical site of localization will affect the amount of the immune response. On the other hand, numerous subclones arise during the development of metastatic disease, and it should be discussed whether and how differences between subclones can be analyzed prior and during vaccination, and how changes of heterogeneity could be integrated into ongoing vaccination.

Question 4: Standardized or Patient-Tailored Vaccines

The immune system of each patient is different. Not only are there genetic and environmental differences prior to onset of the malignancies, but the tumor itself as well as possible pretreatment will have impact on the individual's capacity to mount an anti-tumor immune response. Other factors such as age and lifestyle will further add to the heterogeneity of the immune systems of patients prior to therapy. This heterogeneity is further complicated by the tumor cell variability.

Thus, one has to decide whether standardized or individualized or even continuously optimized vaccines should be chosen. Clearly, standardized vaccines have obvious advantages concerning preparation according to GLP guidelines, and could be produced on a large-scale basis by the pharmaceutical industry. Individualized or even continuously optimized vaccines would offer a better therapeutic response, while having the potential disadvantage of being more time and cost consuming. However, there are techniques available such as *hybrid cell vaccination* that are patient tailored without being cost and time consuming. Before deciding which alternative will be chosen for a given clinical trial, both options should be discussed carefully.

Question 5: Endpoints of Treatment in Vaccination Studies

Metastatic cancer is difficult to cure, with few exceptions. Current treatment options are far from being satisfactory. Therefore, prolongation of overall or disease-free survival, or even improvement in quality of life, are endpoints of most clinical cancer trials.

To study tumor vaccination in metastatic cancer, probably both clinical and immunological endpoints should be chosen. Publications demonstrating the induction of a specific immune response despite lack of clinical response are frequent. This has been regarded as disappointing, but one has to consider that in patients with advanced tumor disease there is virtually no chance to overcome the mass of tumor cells. The demonstration of specific immune responses would indicate that the principle of treatment is in fact promising and should be used in a more appropriate clinical situation.

Question 6: Pros and Cons of Single Agent and Combination Protocols

Immune stimulation by administration of cytokines, such as interleukin 2 or interferon alpha, has been demonstrated to be of therapeutic value, but their use as single therapeutic agents has not been highly satisfactory. Therefore, combination protocols have been designed that combine the advantages of different cytokines or that of chemotherapy and immune activation. In some diseases, such combinations have yielded better results than either treatment modality alone. It is still a matter of debate, at what time point combination treatments of tumor vaccines with surgery, radiation, or chemotherapy should be placed, and what should be the rationale. In the case of high-risk patients with bulky tumor masses, the effects of vaccination may not to be demonstrated because of immune suppression. After surgery, such patients might benefit from tumor vaccination in an adjuvant situation.

Question 7: Can We Develop Information Networks for Tumor Vaccination?

Many other questions remain open. What are the criteria for planning adjuvant studies? Should one perform trials without control groups, or should all studies be conducted with proper control groups?

It seems to me that the situation in immune therapy of cancer resembles that in HIV infection, where there was a tendency to avoid large scale "classic" trials in early phases of drug development. In contrast, small and focused studies have been performed and, in the case of negative results, much time has been saved. The same could be established in tumor vaccination. This would require a highly transparent situation regarding study disclosure. The scientific community should be aware of all studies conducted at a certain time point, and results of studies need to be communicated prior to formal publication. In my view, this will be possible in smaller scientific groups, for example those active in research of certain tumors such as renal cell carcinoma or malignant melanoma. Senior researchers should encourage the establishment of information networks that need to be updated regularly, and meetings of such groups might help to discuss and promote concepts.

Such efforts will be particularly successful if multiple centers agree to perform identical investigations concerning various immune parameters along with the clinical trials. Such data would allow to define criteria that are associated with favorable response to immune therapies.

1.3 Outlook

In summary, many problems remain to be answered in vaccination therapy for malignant diseases. It is important to identify clearly such areas, and to develop strategies to solve open questions. At a time of increasing activities in tumor vaccination, systems need to be established that allow direct access to all relevant knowledge, particularly in the field of ongoing studies. This chapter pleads for cooperative networks which can keep all scientists updated with ongoing studies and their accrual status, and formalized meetings to discuss and promote new concepts.

2 Tumor Antigens

P. Walden

2.1 Introduction

"*In large long-lived animals, like most of the warm-blooded vertebrates, inheritable genetic changes must be common in somatic cells and a proportion of these changes will represent a step towards malignancy. It is an evolutionary necessity that there should be some mechanisms for eliminating or inactivating such potentially dangerous mutant cells ... The thymus-dependent system of immunocytes will be almost solely responsible for surveillance ...*" (Burnet 1970)

With the immune surveillance hypothesis, Macfarlane Burnet summed up about one hundred years of development in tumor immunology. He collated evidence from numerous published reports and personal observations that document the crucial role of the immune system in the defense against cancer. This evidence has been complemented by new investigations and includes postmortem examinations that reveal a high incidence of malignancies that did not become clinically manifest, the demonstration of tumor infiltrating T lymphocytes in many malignancies which often are correlated with a favorable prognosis, spontaneous regression of tumors or tumor regression correlated with the induction of inflammation in the tumors, an increase of the incidence of cancer in patients who suffer from immune deficiency diseases, and, in experimental models, the increased frequency of cancer in thymectomized mice.

Over the past two decades surveillance systems, in addition to the immune system, have been found to participate in the identification and eradication of aberrant cells or at least in the control of their development. These systems act at the level of single cells such as the tumor suppressor factors or at the level of tissues and organs such as angiogenesis and tissue-specific survival factors. However, and in despite of all the surveillance mechanisms, cancer exists as one of the major health problems of mankind and, thereby, testifies to the failure of these systems. Tumor cells in cancer, by their mere existence, document that there are ways to bypass the control instances of the body. These cells have accumulated defects in the genes coding for the molecules that are instrumental in surveillance and, selected for effective escape mechanisms, are highly adaptable. As adaptation implies selection from a diversified tumor cell population, genetic instability and heterogeneity are among the key features of tumor cells, and ongoing selection pressure imposed by the surveillance systems is expected to further diversification. In comparison to the other surveillance systems, the immune system still seems to be best equipped to cope with this heterogeneity and to adapt to changes in the tumor cells. In addition, the immune system, with a few exceptions such as brain and testis, can scan all parts of the body for aberrant cells. Recent publications of the first long-term studies with kidney transplant patients provide clear evidence for the efficiency of immune surveillance. These patients who, in the context of the organ transplantation, had been subjected temporally to immune suppressive treatments have an up to 200-fold increased risk to be affflicted by cancer when compared to age- and sex-matched control individuals (London et al. 1995).

As already suggested by Burnet, thymus-dependent immunocytes, i.e., T cells, are the most important effectors in immune surveillance. With the identification of an increasing number of tumor-associated T cell epitopes during the past 10 or so years, it has become possible to investigate the specific interrelationship of tumor and immune system, to follow the evolution of mutual adaptation, to devise specific vaccines for immune therapy, and to monitor the tumor-specific immunological effects of therapeutic interventions. The following shall give a brief overview of the current status of our knowledge of tumor antigenicity in human cancer, of the currently available technologies for the identification of tumor-associated T cell epitopes, and of the use of the knowledge of such antigens for immune therapy.

2.2 Immune Effector Systems and Tumor Antigenicity

The immune system comprises various effector mechanisms and systems with cytotoxic and, thus, tumoricidal capacity, both, in its innate and its adaptive branch. Many serological tumor markers are known and some are used in cancer diagnostics. However, although antibodies, which are the major instruments of specific serological immune responses, can act as adapters to recruit and activate serological as well as cellular cytotoxic effector mechanisms such as complement and antibody-mediated cellular cytotoxicity with a number of different effector cells and, thus, could participate in the destruction of tumors, very little is known to date about their role in natural anti-tumor defense. Nonetheless, antibodies are being tested in combination with different effector mechanisms such as T cells, complement, cytokines, or toxins for use in cancer immune therapy. Natural killer cells can have their share in immune surveillance as well and, in principle, could complement the activity of MHC restricted $CD8^+$ T cells with their capacity to sense the absence of MHC class I molecules. Loss of MHC expression is one of the major escape mechanisms in human cancer. Again, little is known about the role of these cells in the body of cancer patients. To our current knowledge, $CD8^+$ cytotoxic T cells are the most important effector cells for the defense against cancer. These immunocytes recognize fragments of antigens, peptides of usually nine, in some cases more, amino acids presented by MHC class I molecules at the surfaces of the cells. The MHC restricted mode of antigen recognition enables the T cell to scan internal as well as exported protein antigens and thereby inspect the entire cell for alterations that might be indicative of aberrant developments. At the same time, due to its extreme polymorphism, the MHC causes a high degree of individualization of T cell-mediated immune responses and, thus, of tumor immunology.

Peptides presented by MHC class I molecules are mostly generated from proteins in the cytoplasm by proteolytic enzymes, most notably by proteasomes. They are then transported into the endoplasmic reticulum by specialized peptide transporters, the transporters associated with antigen processing, and there incorporated in nascent MHC class I molecules. Different chaperones participate in these processes in the cytoplasm as well as in the endoplasmic reticulum as intermediate carriers, and additional proteases might trim the peptides further so that they fit into the peptide binding grooves of the MHC molecules. MHC peptide complexes are then exported for display at the cell surface and recognition by the antigen receptor of the T cells. It is the combination of this series of different components in antigen processing and antigen recognition that, each with a different degree of selectivity, ensures the high specificity of T cell-mediated immune responses. Changes in the constitution of the source proteins for the epitopes as well as changes in the activities of components of the antigen processing machinery will have their imprint on the type and quantities of peptides presented by the MHC molecules and thereby on the antigenicity of the tumor cells.

2.3 Techniques for the Identification of Tumor-Associated T Cell Epitopes

The identification of tumor-associated antigens and T cell epitopes is still a major challenge and only a relatively small number of epitopes for a restricted number of tumors and even more restricted immunogenetics, meaning MHC allomorph restriction, are known. The basic requirement for the identification of tumor-associated T cell epitopes is twofold; first, tumor material is needed as the source for the tumor-associated antigens and, second, tumor-specific T cells as the indicator to establish the tumor association and immunological relevance of the identified epitopes. Usually, neither of these two elements is available at the beginning of such work and establishing these requirements is hampered by their mutual dependence, i.e., to identify tumor-associated antigens, tumor-specific T cells are needed and tumor-specific T cells can only be established when tumor cells or peptide analogues of tumor-associated T cell epitopes are available.

Current strategies for the identification of tumor-associated T cell epitopes employ combinations of cellular immunology, molecular genetics, and protein and peptide chemistry to cope with the problems. These strategies include, first, direct isolation and sequencing of MHC-bound peptides by means of high resolution HPLC and mass spectrometry, second, preparation of cDNA expression libraries from tumor cells and transfection of the cDNA into COS cells together with the genes coding for the MHC molecules of the patient from whom the tumor cells were prepared, and subsequent identification of tumor antigen expressing cDNA clones with the patient's CTL (De Plaen et al. 1997; Szikora et al. 1993), third, epitope mapping of known tumor antigens by overlapping peptides or peptides which are predicted to bind to the patients MHC molecules (Kern et al. 1998), and fourth, combinations of the latter two approaches as with SEREX (Serological Screening of Recombinant cDNA Expression Libraries; Sahin et al. 1997) which primarily aims at identifying serologically defined tumor-associated antigens in cDNA expression libraries. These serologically defined tumor-associated antigens are subsequently scanned for potential T cell epitopes (Tureci et al. 1996). Each of these approaches is severely burdened by different problems. For the first, a large amount of tumor material is required which rarely is available. The molecular genetic approach is based on complex transfection schemes, requires several selection steps, and is laborious and time consuming. The third requires prior knowledge of the protein sequence for the tumor-associated T cell epitope. In all cases, tumor-specific T cell clones are needed that are difficult to obtain and that usually suffer from a limited stability in culture. Success with the fourth approach depends on the lucky coincidence of MHC class I and class II restricted T cell epitopes and B cell epitopes on the same polypeptide represented by the serologically determined cDNA. As a consequence of these limitations and despite intense efforts, only relatively few tumor-associated antigens have been identified to date (Boon et al. 1994; van den Eynde and Boon 1997).

Recent developments in combinatorial peptide chemistry offer new options for T cell epitope determination and, thus, also for tumor immunology (Sparbier and Walden 1999). Based on the key features of MHC class I restricted T cell epitopes, randomized peptide libraries can be designed and used for a positional scanning approach to identify the critical amino acids required for the design of the epitope. Epitopes for cytotoxic T cells are usually nonapeptides that conform to MHC allele-specific sequence motifs, i.e., they bear amino acids of restricted variability at sequence positions that anchor the peptide into the peptide-binding groove of the MHC molecules. These motif amino acids vary with the MHC allomorph. The combinatorial peptide libraries and sub-libraries are designed around these features and consist of a mixture of peptides with either all sequence positions randomized or one defined sequence position in the context of randomized positions. A complete set of peptide sub-libraries includes all proteinogenic amino acids and isolates their antigenic properties by combination with randomized positions. This strategy for T cell epitope determination comprises three steps (Gundlach et al. 1996):

1. Identification of the critical amino acids for the induction of a T cell response for every sequence position by testing the responses of a T cell clone to the singly defined combinatorial peptide sub-libraries. This scan yields potential epitope amino acids which are critical for recognition by the T cell receptor.
2. Design and synthesis of potential epitopes that include these critical amino acids as well as the MHC allele-specific motif residues.
3. Identification of the active epitopes in a second T cell assay.

The combinatorial peptide library approach requires no tumor material and is time efficient so that short-term T cell clones will suffice. The epitopes thus determined, however, are not necessarily identical with the natural tumor-associated epitopes. In fact, usually they are not. These so-called mimotopes, mimics of natural epitopes, are often more potent than their natural counterparts. They address slightly different T cell repertoires and thereby are expected to recruit and utilize T cell clones that have not been inactivated by contact with the tumor. Mimotopes have been determined for cutaneous T cell lymphomas and mimotope vaccination protocols are being tested for their capacity to induce anti-tumor immune responses (Linnemann et al. 1998).

2.4 Tumor Associated Antigens

Tumor-associated antigens can be grouped into four categories:

- Viral antigens, antigens encoded by the genome of tumor-associated viruses
- Tumor-specific mutations, point mutations, neo-antigens created by altered posttranslational modifications, translocation sequences, or the sequences resulting from the recombination of immunoglobulin or T cell receptor genes
- Tissue-specific proteins, proteins characteristic for the cell lineage from which the tumor cells descended
- Differential stage-specific proteins, proteins characteristic for earlier differentiation stages of the tumor cell lineage type ontology which are usually expressed in association with germ line or in embryonic cells.

This list reflects the current understanding of tumor-associated antigens. Only the former two of these categories are tumor specific and occur exclusively in the tumor cells; proteins of the latter two groups are expressed in unaltered form also in normal, non-transformed cells of the body and, thus, are regular self-antigens.

Some tumors are associated with carcinogenic viruses. Occasionally such viruses, most notably viruses belonging to the herpes virus family are also found in healthy individuals where they can persist for long periods. These virus-bearing individuals show no clinical signs of cancer and in the vast majority of cases will never develop cancer. Their peripheral blood carries high titers of tumor virus specific cytotoxic T cells and it seems that the immune system is able to control the viruses and prevent cancer without ever achieving sterile immunity. Several virus-specific T cell epitopes are known and it can be assumed that the viruses, usually large DNA viruses, bear a variety of potential T cell epitopes that can be targets for specific T cells. Analyses of tumors developing in immune-suppressed individuals, for example, in cases of HIV infection or therapeutic immune suppression, have revealed that a high proportion of the tumors are associated with tumor viruses (Heeney et al. 1999). These observations suggest that the immune system can keep the viruses under control and prevent cancer in immune-competent individuals without being able to eliminate the virus or the virus-bearing cells. On the one hand, these cases document the high efficiency of the immune system in cancer surveillance and, on the other hand, they seem to indicate that carcinogenic viruses and the immune system can coexist in a state of chronic but controlled infection.

Tumor-specific mutations are ideal target structures for the immune system as no autoreactivity should be induced. The antigen receptors of lymphocytes, immunoglobulins for B cells and T cell receptors for T cells, contain in their junctional regions unique sequences that are only expressed in a single cell clone. All lymphomas thus bear their clone- and, thereby, tumor-specific antigen (Berger et al. 1998). These antigens have been and are being widely used in clinical anti-idiotype vaccination trials for immune therapy of B cell lymphomas. However, these studies have revealed some problems that limit the efficiency and applicability of anti-idiotype vaccination therapy. A relatively high frequency of mutations has been reported for the immunoglobulin genes of B cell lymphomas and loss of expression of T cell receptors in some cases of T cell lymphoma (Forste et al. 1997; Gellrich et al. 1997). Such modifications of the structure or expression of the putative tumor antigens can be interpreted as an effect of immune selection which would imply that these sequences are indeed used by the immune system to target the tumor cells. In B cell lymphomas, diversification of the immunoglobulin genes is particularly pronounced after anti-idiotype vaccination. Also other genes may show high mutation rates in tumor cells. A series of mutations has been identified in the tumor suppresser gene p53, some of which lead to T cell epitopes. A particularly interesting case of a tumor-specific mutation has been reported for CDK4, a protein involved in cell cycle control and it has been speculated that this mutation, which gives rise to a T cell epitope, might be especially interesting as CDK4 is believed to be important for the malignancy of the tumor cells in this particular cancer (Wölfel et al. 1995). While being very promising, utilization of the tumor-specific mutations is hampered by the fact that they are, in most cases, restricted to an individual tumor in a single patient. The only exception identified so far is a neo-antigen generated by the bcr/abl translocation which occurs in chronic myeloid leukemia and which results in the expression of unique T cell epitopes (Yasukawa et al. 1998).

Examples for tissue type-specific tumor-associated antigens are tyrosinase, gp100, and gp75 which are expressed in all pigmented cells including melanocytes and many melanoma cells (Kawakami et al. 1998). These antigens have been identified with the aid of tumor-specific T cells obtained from cancer patients. It must, therefore, be concluded that autospecific T cells are present in the patients which target self-antigens. In fact, autoimmune responses against pigmented cells have been seen as vitiligo occurring in melanoma patients but also in healthy individuals (Rosenberg and White 1996). The existence of T cells with specificity for normal tissue antigens in these patients raises basic questions with respect to the mechanisms of self-tolerance, questions which have not been resolved yet. Obviously, these autospecific cells were not clonally deleted by negative selection in the thymus. Vitiligo in the context of anti-melanoma immune responses shows that there is a fine balance between immune tolerance of self, anti-tumor immunity, and autoimmune reactivity. Frequent reports about vitiligo in melanoma patients who respond to vaccination therapy with the above-mentioned antigens also illustrate the capacity of the immune system to utilize self-antigens to target and destroy the tumor cells. Despite the reports of cancer-associated autoimmune reactions in melanoma patients and especially in patients vaccinated against tumor-associated self-antigens, there is no report about an autoimmune disease induced by therapeutic vaccination.

A large fraction of the known tumor-associated antigens that have been identified in recent years were found to be expressed in melanoma cells and in testis but in no other tissue (van den Eynde and Boon 1997; De Smet et al. 1997). Testis is an immune-privileged organ protected from the immune system. MAGE, BAGE, and GAGE belong to this group of tumor-associated antigens. So far no sequence relationship or structural similarity with known protein families could be found for these antigens. There is also no indication of the physiological function of these proteins. However, they are expressed in a high frequency of melanomas, and T cells with specificity for these antigens can be demonstrated in the peripheral blood of many patients and even in some healthy individuals (Chaux et al. 1998). Recently, some of the melanoma-associated, testis-specific antigens have been found in other tumors as well (Weynants et al. 1994). MAGE, for instance, is expressed in many cases of head and neck cancer and in a number of cases of hematological malignancies. These antigens could be examples of a group of widely shared antigens associated with the general pathophysiology of the tumor.

The vast majority of the tumor-associated T cell epitopes identified up to now are derived from proteins that are not essential for survival or for the malignant properties of the tumor cells. They can do well without these proteins and many examples have been reported where these antigens were lost without a detectable interference with tumor progression. On the contrary, at more advanced stages of cancer an increasing frequency of antigen loss variants of the tumor cells can be detected. While antigen loss can be, and most likely is, a stochastic process occurring without any involvement of the immune system, it can be assumed that the selection pressure imposed by anti-tumor immune responses is the major factor in selecting these antigen loss variants. As has been documented for other areas of biology, the combination of a high mutation rate and vigorous selection pressure can be the driving force for an increasing degree of heterogenicity in the tumor cell populations which thus would evade immune surveillance. Such prospects emphasize that it is important to identify as many tumor-associated antigens as possible which can be used to vaccinate simultaneously against as many tumor variant cells as possible. Vaccination schemes that address complex arrays of tumor-associated antigens are expected to be more effective in many cases than single antigen vaccines. The above considerations would also imply that vaccination therapy should be applied as early as possible in the development of the disease to hit the tumor before antigenic diversification and antigen losses render the actions of the immune system futile.

2.5 Tumor Immunogenicity

While antigenicity defines the property of being recognizable by the immune system, i.e., by antibodies and T cells, immunogenicity implies the capacity of the immunogen to induce immune responses. The induction of cytotoxic effector T cells, just like the induction of specific humoral immune responses, requires T cell help (Borges et al. 1994). The basic regulatory unit in the case of cytotoxic T cell responses is a three-cell cluster composed of an antigen-presenting cell (APC), a helper T cell, and a precursor of a cytotoxic T cell (Stuhler and Walden 1993; Brunner et al. 1994). This situation differs from the two-cell cluster of T and B cells that controls the induction and development of specific antibody-mediated immunity. The APC is the organizer of this regulatory three-cell cluster system and has to present on its MHC class I and class II molecules epitopes for the cytotoxic precursor cell and the helper T cell, respectively (Stuhler and Walden 1993). Thus, there is a twofold MHC restriction that controls the induction of cytotoxic T cell responses (Juretic et al. 1985) in contrast to T-B cell collaboration and antibody formation which is controlled solely by MHC class II molecules. In both cases, the two epitopes for the two interacting antigen-specific lymphocytes must be physically linked; however, the physical basis for epitope linkage in the case of T-T cell collaboration is a cell, the APC, not a molecule as it is the case for T-B cell collaboration. The twofold MHC restriction control for the induction of cytolytic effector T cell responses provides additional protection against potential autoimmune reactions even though the antigens for the two interacting T cells need not be related on the molecular level. The involvement of two different MHC molecules in the induction of effector T cell responses, the requirement for suitable tumor-associated antigens which bear epitopes that can be presented by these MHC molecules, and the need for tumor-specific T cells that must be present in the T cell repertoire of the patient result in the known high degree of individualization of anti-cancer immunity. While most tumor cells express MHC class I molecules, there are only a few examples of MHC class II expression. Nonetheless, recently the first MHC class II restricted tumor-associated T cell epitopes have been identified by scanning known melanoma-associated antigens for potential epitopes for helper T cells (Hiltbold et al. 1998; Kobayashi et al. 1998; Manici et al. 1999). However, the double MHC restriction requirement still poses a major hurdle for induction of tumor-specific cytotoxic T cells which has to be dealt with in vaccination therapy.

Recognition of cognate peptides together with their specific MHC class I and class II molecules on the surface of the APC by the two T cell types induces a complex array of molecular interaction between these cells. Both T cells communicate with the APC and with each other via cell surface molecules and soluble factors (Stuhler and Walden 1993). These molecules are either constitutively expressed or are induced during the interaction of the cells. Their expression may be a result of the maturation of the participating cells as is the case for CD28, B7, and 4–1BB, or they may be activation induced such as CD40 on the APC and CD40L on the T cells (Ridge et al. 1998; Schoenberger et al. 1998), or CD27 and CD70 which are involved in direct T-T cell interaction (Stuhler et al. 1999). The expression and secretion of the lymphokines interleukin-2, as growth and differentiation factor for CTL provided by the helper T cell, and γ-IFN, a major differentiation factor of CTL which can be produced by both T cell types, are strictly activation induced. These two soluble factors are crucial for communication between the T cells and, under physiological conditions, act over a short distance. All these molecular interactions provide signals for costimulation, modulation, survival, growth, and differentiation of the T cells and are essential for the development of cellular immunity. Still more new players are being discovered or are implicated by indirect evidence and are added to the list.

Although several MHC class II-expressing cells can serve as APCs in the regulatory three-cell interactions, dendritic cells are clearly the most potent cells for antigen presentation, for organization of the regulatory clusters, and for costimulation (Schuler et al. 1997; Banchereau and Steinman 1998). The impact of the specific architecture of the lymphoid organs on the efficiency of the cellular interactions and of the induction of effector T cells is largely unexplored. Dendritic cells harvest antigen from the tissues, become activated by inflammatory processes, and transport the processed antigen to lymphoid organs for induction of immune responses. Non-lymphoid tissues are far less potent promoters of immune responses. In contrast to the multiple regulatory interactions controlling the induction of cytotoxic T cells, execution of the effector functions, i.e., the cytolysis of the target cells, is dependent only on the T cell receptor, MHC class I peptide complexes, and a few adhesion molecules.

2.6 Therapeutic Vaccination for the Treatment of Cancer

Vaccination is usually applied for prevention of infectious diseases by induction of immune responses aimed at long-lasting immunity. In cancer, however, the disease has already invaded the individual and usually has been developing in the patient's body for a prolonged time. Thus, the aim of the vaccination in cancer is foremost the therapy of an ongoing disease which implies modulation of an ongoing immune response rather than initiation of new immune responses. To what extent this difference requires different vaccination strategies is still being investigated. However, the goal of therapeutic vaccination for cancer therapy is a specific shift in the quality and quantity of tumor-specific immune responses which are the result of constant mutual adaptations with the tumor and which are continuously changing in the course of these interactions. Therefore, the development of oncological vaccination strategies must be concerned with both the identification and utilization of suitable target antigens and the systematic and controllable modulation of the immune regulatory context and requirements for the induction of effective anti-tumor cytolytic T cells (Herlyn and Birebent 1999). Various concepts have been suggested to achieve this goal and some of these concepts are being tested in the ward. Crucial for success of these endeavors is the systematic development and improvement of therapeutic vaccination strategies by evaluation of the tumor antigen specific effects of the therapeutic intervention. Only recently new technologies for the detection and accurate enumeration of antigen-specific cells, such as the MHC tetramers techniques (Altman et al. 1996) or intracellular cytokine staining of cells responding to tumor-associated T cell epitopes (Kern et al. 1998), have been developed and, in combination with the gradually increasing number of identified tumor-associated T cell epitopes, have opened new prospects for the development of new therapies for cancer.

References

Altman JD, Moss PAH, Goulder PJR, Barouch DH, McHeyzer Williams MG, Bell JI, McMichael AJ, Davis MM (1996) Phenotypic analysis of antigen-specific T lymphocytes. Science 274:94–96

Banchereau J, Steinman RM (1998) Dendritic cells and the control of immunity. Nature 392:245–252

Berger CL, Longley BJ, Imaeda S, Christensen I, Heald P, Edelson RL (1998) Tumor-specific peptides in cutaneous T-cell lymphoma: association with class I major histocompatibility complex and possible derivation from the clonotypic T-cell receptor. Int J Cancer 76:304–311

Boon T, Cerottini JC, Eynde B van den, Van Der Bruggen P, Van Pel A (1994) Tumor antigens recognized by T lymphocytes. Annu Rev Immunol 12:337–365

Borges E, Wiesmuller KH, Jung G, Walden P (1994) Efficacy of synthetic vaccines in the induction of cytotoxic T lymphocytes. Comparison of the costimulating support provided by helper T cells and lipoamino acid. J Immunol Methods 173:253–263

Brunner MC, Mitchison NA, Schneider SC (1994) Immunoregulation mediated by T-cell clusters. Folia Biol (Praha) 40:359–369

Burnet M (1970) The concept of immune surveillance. Prog Exp Tumor Res 13:1–27

Chaux P, Vantomme V, Coulie P, Boon T, Van Der Bruggen P (1998) Estimation of the frequencies of anti, MAGE, 3 cytolytic T-lymphocyte precursors in blood from individuals without cancer. Int J Cancer 77:538–542

De Plaen E, Lurquin C, Lethe B, Bruggen P van der, Brichard V, Renauld JC, Coulie P, Van Pel A, Boon T (1997) Identification of genes coding for tumor antigens recognized by cytolytic T lymphocytes. Methods Enzymol 12:125–142

De Smet C, Martelange V, Lucas S, Brasseur F, Lurquin C, Boon T (1997) Identification of human testis, specific transcripts and analysis of their expression in tumor cells. Biochem Biophys Res Commun 241:653–657

Eynde BJ van den, Boon T (1997) Tumor antigens recognized by T lymphocytes. Int J Clin Lab Res 27:81–86

Forste N, Gellrich S, Golembowski S, Rutz S, Audring H, Sterry W, Jahn S (1997) Analysis of V(H) genes rearranged by individual B cells in dermal infiltrates of patients with mycosis fungoides. Clin Exp Immunol 110:464–471

Gellrich S, Golembowski S, Audring H, Jahn S, Sterry W (1997) Molecular analysis of the immunoglobulin VH gene rearrangement in a primary cutaneous immunoblastic B-cell lymphoma by micromanipulation and single, cell PCR. J Invest Dermatol 109:541–545

Gundlach BR, Wiesmuller KH, Junt T, Kienle S, Jung G, Walden P (1996) Determination of T cell epitopes with random peptide libraries. J Immunol Methods 192:149–155

Heeney JL, Beverley P, McMichael A, Shearer G, Strominger J, Wahren B, Weber J, Gotch F (1999) Immune correlates of protection from HIV and AIDS: more answers but yet more questions. Immunol Today 20:247–251

Herlyn D, Birebent B (1999) Advances in cancer vaccine development. Ann Med 31:66–78

Hiltbold EM, Ciborowski P, Finn OJ (1998) Naturally processed class II epitope from the tumor antigen MUC1 primes human $CD4^+$ T cells. Cancer Res 58:5066–5070

Juretic A, Malenica B, Juretic E, Klein J, Nagy ZA (1985) Helper effects required during in vivo priming for a cytolytic response to the H-Y antigen in nonresponder mice. J Immunol 134:1408–1414

Kawakami Y, Robbins PF, Wang RF, Parkhurst M, Kang X, Rosenberg SA (1998) The use of melanosomal proteins in the immunotherapy of melanoma. J Immunother 21:237–246

Kern F, Surel IP, Brock C, Freistedt B, Radtke H, Scheffold A, Blasczyk R, Reinke P, Schneider-Mergener J, Radbruch A, Walden P, Volk HD (1998) T cell epitope mapping by flow cytometry. Nat Med 4:975–978

Kobayashi H, Kokubo T, Sato K, Kimura S, Asano K, Takahashi H, Iizuka H, Miyokawa N, Katagiri M (1998) $CD4^+$ T cells from peripheral blood of a melanoma patient recognize peptides derived from nonmutated tyrosinase. Cancer Res 58:296–301

Linnemann T, Brock C, Sparbier K, Muche M, Mielke A, Lukowsky A, Sterry W, Kaltoft K, Wiesmuller KH, Walden P (1998) Identification of epitopes for CTCL, specific cytotoxic T lymphocytes. Adv Exp Med Biol 451:231–235

London NJ, Farmery SM, Will EJ, Davison AM, Lodge JP (1995) Risk of neoplasia in renal transplant patients. Lancet 346:403–406

Manici S, Sturniolo T, Imro MA, Hammer J, Sinigaglia F, Noppen C, Spagnoli G, Mazzi B, Bellone M, Dellabona P, Protti MP (1999) Melanoma cells present a MAGE-3 epitope to $CD4^+$ cytotoxic T cells in association with histocompatibility leukocyte antigen DR11. J Exp Med 189:871–876

Ridge JP, Di Rosa F, Matzinger P (1998) A conditioned dendritic cell can be a temporal bridge between a $CD4^+$ T-helper and a T-killer cell. Nature 393: 474–478

Rosenberg SA, White DE (1996) Vitiligo in patients with melanoma: normal tissue antigens can be targets for cancer immunotherapy. J Immunother Emphasis Tumor Immunol 19:81–84

Sahin U, Tureci O, Pfreundschuh M (1997) Serological identification of human tumor antigens. Curr Opin Immunol 9:709–716

Schoenberger SP, Toes RE, Van Der Voort EI, Offringa R, Melief CJ (1998) T-cell help for cytotoxic T lymphocytes is mediated by CD40-CD40L interactions. Nature 393:480–483

Schuler G, Thurner B, Romani N (1997) Dendritic cells: from ignored cells to major players in T-cell, mediated immunity. Int Arch Allergy Immunol 112:317–322

Sparbier K, Walden P (1999) T cell receptor specificity and mimotopes. Curr Opin Immunol 11:214–218

Stuhler G, Walden P (1993) Collaboration of helper and cytotoxic T lymphocytes. Eur J Immunol 23:2279–2286

Stuhler G, Zobywalski A, Grunebach F, Brossart P, Reichardt VL, Barth H, Stevanovic S, Brugger W, Kanz L, Schlossman SF (1999) Immune regulatory loops determine productive interactions within human T lymphocyte, dendritic cell clusters. Proc Natl Acad Sci USA 96:1532–1535

Szikora JP, Van Pel A, Boon T (1993) Tum^{-} mutation P35B generates the MHC-binding site of a new antigenic peptide. Immunogenetics 37:135–138

Tureci O, Sahin U, Schobert I, Koslowski M, Scmitt H, Schild HJ, Stenner F, Seitz G, Rammensee HG, Pfreundschuh M (1996) The SSX-2 gene, which is involved in the t(X,18) translocation of synovial sarcomas codes for the human tumor antigen HOM-MEL-40. Cancer Res 56:4766–4772

Weynants P, Lethe B, Brasseur F, Marchand M, Boon T (1994) Expression of MAGE genes by non-small-cell lung carcinomas. Int J Cancer 56:826–829

Wölfel T, Hauer M, Schneider J, Serrano M, Wölfel C, Klehmann-Hieb E, De Plaen E, Hankeln T, Meyer-zum-Buschenfelde KH, Beach D (1995) A p16INK4a-insensitive CDK4 mutant targeted by cytolytic T lymphocytes in a human melanoma. Science 269:1281–1284

Yasukawa M, Ohminami H, Kaneko S, Yakushijin Y, Nishimura Y, Inokuchi K, Miyakuni T, Nakao S, Kishi K, Kubonishi I, Dan K, Fujita S (1998) CD4^{+} cytotoxic T-cell clones specific for bcr-abl b3a2 fusion peptide augment colony formation by chronic myelogenous leukemia cells in a b3a2-specific and HLA-DR-restricted manner. Blood 92:3355–3361

3 Costimulatory Molecules in Vaccine Design

J. W. Hodge, J. Schlom

3.1 Introduction

T cell activation has now been shown to require at least two signals. The first signal is antigen specific, is delivered through the T cell receptor via the peptide/MHC complex, and causes the T cell to enter the cell cycle. The second, or costimulatory, signal is required for cytokine production and proliferation, and is mediated through ligand interaction on the surface of the T cell. Several molecules normally found on the surface of professional antigen-presenting cells (APC) have been shown to be capable of providing the second signal critical for T cell activation. These molecules include B7–1 (CD80), B7–2 (CD86), intercellular adhesion molecule-1 (ICAM-1, CD54), and leukocyte function-associated antigen-3 (LFA-3, human CD58/murine CD48), among others. T cell costimulation has now been shown to be delivered via several modalities and delivery systems (i.e., recombinant retroviral vectors, recombinant poxviral vectors, and anti-CTLA-4 antibodies) to enhance anti-tumor immunity in experimental models. This chapter will deal principally with the use of recombinant poxvirus vectors to deliver an array of costimulatory molecules, via either direct vaccination approaches or whole-tumor-cell vaccines, to induce anti-tumor immunity. Both prevention and tumor-therapy models will be discussed.

Specifically, this chapter will deal with: (a) comparative studies on the use of a dual-gene construct of a recombinant vaccinia (rV) vector containing a tumor-associated antigen (TAA) gene and a costimulatory molecule gene versus the use of admixtures of the replication-competent rV-TAA and recombinant vaccinia containing the costimulatory molecule to induce anti-tumor immunity, (b) the use of an admixture of vaccinia viruses containing a TAA gene and the B7–1 costimulatory molecule gene to induce a therapeutic response in a lung metastasis tumor model, (c) the anti-tumor efficacy of whole-tumor-cell vaccines in which the B7–1 costimulatory molecule is expressed in a tumor-cell vaccine via a vaccinia versus a retroviral vector, (d) the use of recombinant poxviruses containing the genes for the costimulatory molecules ICAM-1 or LFA-3 to induce anti-tumor immunity, and (e) the use of poxvirus vectors containing a triad of costimulatory molecules (B7–1, ICAM-1, and LFA-3, designated TRICOM; Hodge et al. 1999a) that synergize to enhance both $CD4^+$ and $CD8^+$ T cell responses to a new threshold. Some of the review material presented here has been discussed previously.

3.2 Poxvirus Vaccines: Potential Advantages and Disadvantages

Recombinant poxviruses have been used in a wide range of vaccines in experimental and clinical studies. Most of these vaccines have been directed against viral antigens such as rabies and HIV (Graham et al. 1992; Egan et al. 1995; Fries et al. 1996; Tubiana et al. 1997; Belshe et al. 1998; Clements-Mann et al. 1998). Thus, both rodent models and clinical studies have revealed many of the advantages and disadvantages of recombinant poxvirus vectors.

Vaccination with a live recombinant vaccinia virus allows for the expression of foreign antigens encoded by a transgene directly in various cells of the host, including professional APC. This method of vaccination enables antigen processing and presentation of antigenic peptides along with host histocompatibility antigens and other necessary cofactors found on the APC. That these foreign antigens are presented to the immune system with a large number of proteins produced by the vector itself most likely accounts for the significant inflammatory response to the poxvirus vector. In turn, this inflammatory process apparently leads to an environment of cytokine production and T cell proliferation that may further amplify the immune response to the foreign antigen. This process favors induction of a cell-mediated immune response and humoral responses to the foreign antigen. Because vaccinia actively replicates in the host, it can present high levels of antigen to the immune system over a period of 1–2 weeks, substantially increasing the potential for immune stimulation. The host immune response to the vaccinia vector then eliminates the virus. While this scenario is excellent for inducing immune responses to bystander transgene products, this same phenomenon also limits the number of administrations of recombinant vaccinia vectors. After one or two administrations of recombinant vaccinia, the host mounts potent anti-vaccinia antibody and T cell responses (Demkowicz et al. 1996; Stienlauf et al. 1999). This reduces vaccinia's ability to replicate in subsequent booster vaccines and, hence, leads to limited transgene expression. However, recent clinical studies have shown that the administration of one or two vaccinations of a recombinant vaccinia virus can elicit T cell responses to the inserted transgene even in patients who previously received the smallpox vaccine (McAneny et al. 1996; Sanda et al. 1999). Several experimental studies and recent clinical studies have also demonstrated that recombinant vaccinia viruses are best used for priming the immune response (Bei et al. 1994; Hodge et al. 1997; Marshall et al. 1999). Subsequent vaccinations can employ proteins, peptides, DNA vaccines, or replication-defective poxviruses, as well as other recombinant vectors (Montefiori et al. 1992; Bei et al. 1994; Graham et al. 1994; Cole et al. 1996; Hodge et al. 1997; Marshall et al. 1999).

Modified Vaccine Ankara (MVA) is a replication-defective poxvirus derived from vaccinia following 500 passages in chicken embryo cells. It has been employed in many experimental studies and has now been administered to more than 120,000 people without apparent side effects (Moss 1996). Recently, this virus has been molecularly characterized and has been found to have lost several genes involved in host-range determination and possible immune system suppression. While MVA efficiently infects human cells and expresses both early and late genes, it is replication-defective and incapable of producing infectious progeny in mammalian cells. MVA recombinant viruses have been shown to be highly immunogenic in both rodents and primates. To date, only a few experimental studies have employed recombinant MVA as anti-cancer vaccines (Moss et al. 1996; Carroll et al. 1997).

Avipox viruses such as fowlpox and canarypox (ALVAC) are nonreplicating and thus represent potentially attractive vectors for use in cancer vaccines. While the immunogenicity of the inserted transgene may not be as potent as that of vaccinia virus, avipox viruses can be administered numerous times to enhance immunogenicity (Fries et al. 1996; Hodge et al. 1997; Tartaglia et al. 1998). Since they are replication-defective, induction of any host immune responses should be inconsequential. Avipox viruses are also distinguished from vaccinia in that the inserted transgene is expressed in infected cells for 14–21 days prior to the death of the cell. In a vaccinia-infected cell, the transgene is expressed for 1–2 days until cell lysis, and for approximately 1 week in the host until virus replication is arrested by host immune responses.

One of the advantages of using recombinant poxviruses as anti-cancer vaccines is the ability to insert large amounts of foreign DNA and multiple genes. To date, as many as seven genes have been inserted into vaccinia virus (Ockenhouse et al. 1998). Generally, poxvirus-based vaccines have been shown to be cost-effective, safe, easy to administer, and stable for long periods of time without special storage conditions. Other advantages include: (a) a wide host range and cell-type range, (b) stability, (c) accurate replication, (d) efficient post-translational processing of the inserted transgene, and (e) the tendency of recombinant gene products to be more immunogenic.

A number of investigators have also demonstrated the advantage of priming with vaccinia recombinants and boosting with immunogens such as recombinant protein, peptide, or recombinant fowlpox or avipox vectors to enhance immune responses in cancer vaccine models (Kahn et al. 1991; Bei et al. 1994; Cole et al. 1996; Hodge et al. 1997). The experimental studies that involved priming with recombinant vaccinia and boosting with recombinant avipox also demonstrated that the host immune response to the transgene increased with continued booster vaccinations (Hodge et al. 1997).

3.3 Enhancing Antigen-Specific T Cell Immunity via T Cell Costimulation: the Use of Combination Vaccinia Vaccines and Dual-Gene Vaccinia Vaccines

Recombinant vaccinia viruses expressing costimulatory transgenes have recently been shown to enhance T cell responses to antigen-specific vaccines. This has been accomplished using two different modalities. The first modality employs the use of combination vaccines. For example, rV-carcinoembryonic antigen (CEA) was admixed with rV-B7–1 (murine) and was shown to elicit enhanced T cell responses and anti-tumor activity in a murine model (Hodge et al. 1995). The second modality to express costimulatory molecules in vaccinia viruses involves the use of dual-gene vaccines (Kalus et al. 1999; i.e., in which both the target gene and the gene encoding the costimulatory molecule are inserted into the same vaccinia virus genome). There are potential advantages in using either one of these modalities to enhance antigen-specific responses via costimulation. Using combination vaccines, one can employ multiple combinations of antigen-specific recombinant viruses admixed with a given costimulatory recombinant virus. For example, rV-B7–1 can be admixed with either rV-CEA, rV-MUC-1, rV-PSA, or other recombinant viruses expressing a gene encoding a target antigen. Conversely, one can admix a particular recombinant virus containing the gene for a specific antigen with one of a range of recombinant vaccinia viruses containing different costimulatory molecules, such as rV-B7–1, rV-B7–2, rV-ICAM-1, rV-LFA-3, and rV-CD70, etc. The cost- and time-effectiveness of this approach is clearly advantageous both in experimental studies and, particularly, in the preparation, quality control, and regulatory considerations of clinical grade reagents. Yet another advantage of the use of combination vaccines is that one can alter the ratios of the two recombinant vaccines employed for optimal effects. The potential advantages of using a dual-gene recombinant virus are: (a) only one reagent is employed and (b) the coexpression of both transgenes in the same cell is guaranteed.

Recent comparative studies on the use of both approaches to enhance antigen-specific T cell responses have shown the importance of route of vaccine administration and vaccine dose in attaining optimal T cell responses (Kalus et al. 1999). These studies should have direct bearing on the design of vaccine clinical trials for infectious agents and/or TAAs, in which T cell costimulatory molecules will be employed to enhance antigen-specific T cell responses via the use of either combination or dual-gene vaccinia vaccines.

In vitro studies were first undertaken to determine if the combined use of rV-CEA with rV-B7–1 can indeed superinfect the same cells and thus express both transgene products simultaneously. The rV-CEA/B7–1 dual-gene vaccine was used as the positive control. When rV-CEA/B7–1 was used to infect BSC-1 cells at a multiplicity of infection (MOI) of 5 and analyzed by two-color flow cytometry analysis, 90% of the cells expressed both transgenes (Fig. 1C). When the combination of rV-CEA and rV-B7–1 was used at the same MOI, however, only 44% of the cells expressed both transgenes (Fig. 1A). When used at an MOI of 10, the dual-gene vaccine was shown to express both transgenes on 84% of cells (Fig. 1D). When the combination of rV-CEA and rV-B7–1 was used at 10 MOI, however, 82% of cells expressed both transgenes simultaneously (Fig. 1B). Thus, as one would expect, the higher MOI resulted in a more efficient simultaneous expression of both genes in infected cells. The route of administration of recombinant vaccinia vaccines was then explored in a murine model. The efficiency

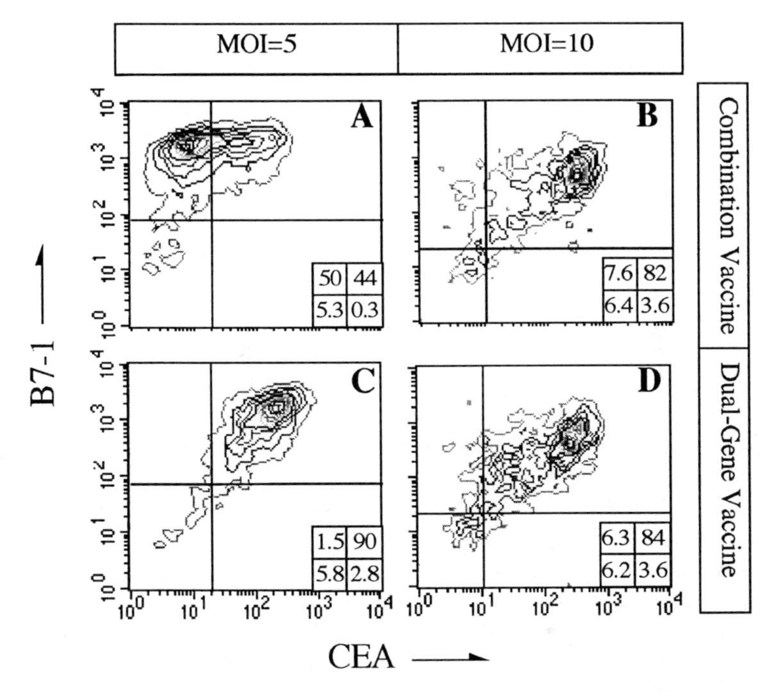

Fig. 1A–D. Fluorescent analysis of surface coexpression of carcinoembryonic antigen (*CEA*) and B7–1 molecules. BSC-1 cells were either co-infected with recombinant vaccinia (rV)-CEA:rV-B7–1 combination vaccine (**A,B**) or infected with the rV-CEA/B7–1 dual-gene vaccine (**C,D**). Cells were infected at a multiplicity of infection (*MOI*) of either 5 (**A,C**) or 10 (**B,D**), and stained 6 h postinfection with monoclonal antibodies directed against murine B7–1 (PE-B7–1) and directed against CEA (FITC-COL-1). Percentages of positive cells in each quadrant are depicted in *inset panels*

of skin scarification (the conventional route of vaccinia administration in humans as well as in most experimental models) was compared to the intravenous route of administration. As seen in Fig. 2A,B, when wild-type vaccinia (V-Wyeth or V-WT) or rV-B7 were administered by either route and splenic lymphocytes were analyzed 14 days later, no proliferative T cell responses to CEA were observed in any of the concentrations used. When rV-CEA was administered via scarification or intravenously, similar ($p>0.05$) proliferative responses to CEA were observed

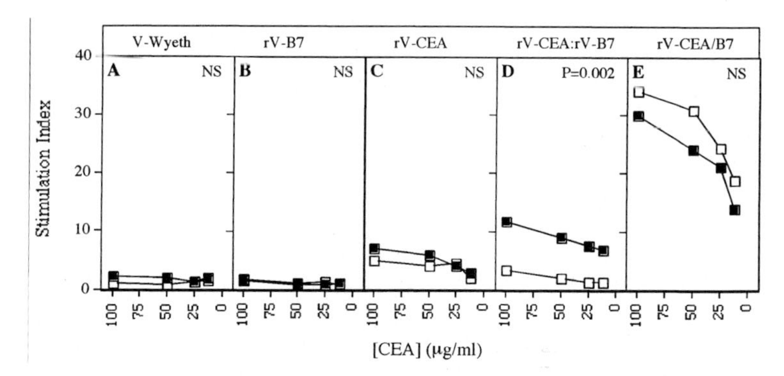

Fig. 2A–E. CEA-specific lymphoproliferative T cell responses following vaccination with rV-CEA:rV-B7 combination vaccine or vaccination with rV-CEA/B7 dual-gene vaccine by two different routes. Mice were vaccinated by tail scarification (*closed squares*) or intravenously (*open squares*) with 1×10^7 pfu of either wild-type vaccinia virus (V-Wyeth; **A**), rV-B7 (**B**), rV-CEA (**C**), rV-CEA:rV-B7 combination vaccine (**D**), or rV-CEA/B7 dual-gene vaccine (**E**). Splenocytes were harvested 14 days after vaccination and analyzed for CEA-specific lymphoproliferation. *P* values were calculated at 95% via ANOVA and depicted in the *upper right hand quadrant* of each panel. Values were compared between routes for the same immunogen. *NS,* Not significant

with either route (Fig. 2C). When the combination vaccine of rV-CEA and rV-B7 was administered intravenously, low levels of CEA-specific T cell proliferation were observed. Statistically significant (p=0.002) differences in T cell proliferation were observed, however, when this combination vaccine was administered via scarification versus the intravenous route (Fig. 2D). No significant differences were observed when the dual-gene vaccine was administered by either route (Fig. 2E). At this dose, 1×10^7 pfu, the dual-gene vaccine elicited stronger (p=0.002) proliferative T cell responses than the combination vaccine via the scarification route (Fig. 2E versus Fig. lD). Experiments demonstrated little if any difference in the induction of anti-tumor immunity when the combination vaccine or the dual-gene vaccine was administered by the subcutaneous route (Fig. 3).

Studies were then undertaken to determine the effect of vaccine dose on T cell responses. Following administration of either the dual-gene

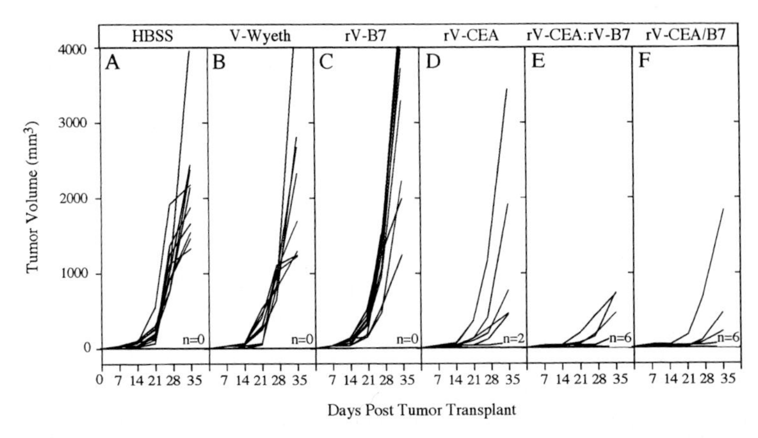

Fig. 3A–F. Growth of transplanted murine adenocarcinoma cells expressing CEA in mice vaccinated with rV-CEA:rV-B7–1 combination vaccine or vaccinated with rV-CEA/B7–1 dual-gene vaccine. Ten C57BL/6 mice/group were ivaccinated by tail scarification with Hanks' balanced salt solution (HBSS; **A**) or with 1×10^7 pfu of either wild-type vaccinia (V-Wyeth; **B**), rV-B7–1 (**C**), rV-CEA (**D**), rV-CEA:rV-B7–1 (**E**), or rV-CEA/B7–1 (**F**), and injected 14 days later with 3×10^5 MC38-CEA-2 murine adenocarcinoma cells expressing CEA. *n*, Number of mice that remained tumor free

vaccine or the combination vaccine via scarification, splenocytes of vaccinated mice were evaluated for their ability to mediate T cell proliferative responses 28 days after vaccination administration. As seen in Table 1, when administered at a dose of 5×10^6 pfu, there was a clear statistical difference (p=0.006) in the advantage of the use of the dual-gene vaccine over the combination vaccine. Similar results were seen using a dose of 1×10^7 pfu (Table 1). However, when the higher dose of 5×10^7 pfu was employed, no statistical difference (p=0.13) was observed when either vaccine was used (i.e., the combination vaccine and the dual-gene vaccine elicited similar T cell responses; Table 1). Similarly, when the combination vaccine or the dual-gene vaccine was administered at the 1×10^8 pfu dose via scarification, no statistical differences (p=0.184) were observed among the two vaccine types in the induction of T cell responses (Table 1). These studies corroborate the in vitro experiments (Fig. 1), demonstrating that when higher doses of

Table 1. The effect of vaccine dose on the efficacy of combination or dual-gene vaccinia vaccines

Immunogen[a]	Dose[b]	Antigen[c]						P value[d]
		Con A	Oval-bumin	CEA (μg/ml)				
				100	50	25	12.5	
HBSS	NA	161	1.0	3.5	2.5	3.0	1.1	
V-WT	1×10^8	154	0.8	2.8	2.7	1.9	2.1	
rV-CEA:rV-B7–1	5×10^6	182	0.4	18.4	11.7	6.7	6.1	
rV-CEA/B7–1	5×10^6	165	0.7	30.1	24.9	16.4	9.0	0.0061
rV-CEA:rV-B7–1	1×10^7	158	1.3	30.2	19.1	8.9	12.1	
rV-CEA/B7–1	1×10^7	150	0.8	50.2	38.1	34.2	29.0	0.0031
rV-CEA:rV-B7–1	5×10^7	147	0.9	49.1	41.5	32.9	22.1	
rV-CEA/B7–1	5×10^7	164	1.2	56.2	46.0	30.1	27.6	NS
rV-CEA:rV-B7–1	1×10^8	139	1.0	51.3	45.3	35.1	30.4	
rV-CEA/B7–1	1×10^8	151	1.1	53.1	51.0	43.5	29.7	NS

CEA, carcinoembryonic antigen; HBSS, Hanks' balanced salt solution; V-WT, wild-type vaccinia; rV, recombinant vaccinia; NA, not applicable; NS, not significant

[a]Five C57BL/6 mice were vaccinated once (scarification) with the indicated combination vaccine or dual-gene vaccine. Lymphoproliferative responses from pooled splenic T cells were analyzed 28 days following vaccination

[b]The dose indicates the total number of pfu per mouse

[c]Antigen concentrations were: Con A (2 μg/ml), ovalbumin (100 μg/ml), and CEA (100–12.5 μg/ml). Each value represents the stimulation index of the mean CPM of triplicate samples versus media

[d]*P* values were calculated at 95% via ANOVA. Values were compared between members of the same dose level

vaccine are employed, the use of combination vaccines and dual-gene vaccines yields statistically similar results.

3.4 Therapeutic Anti-Tumor Response after Vaccination with an Admixture of Recombinant Vaccinia Virus Expressing a Modified MUC-1 Gene and the Murine T Cell Costimulatory Molecule B7–1

DF3/MUC-1 is a TAA that is overexpressed with an abnormal glycosylation pattern in breast, ovarian, lung, and pancreatic cancers. The major extracellular portion of MUC-1 is composed of tandem repeat units of 20 amino acids. Recombinant vaccinia viruses encoding mucin molecules have been constructed by several groups. However, these recombinants have met with limited success in protecting animals from tumors expressing MUC-1 because the vaccinia genome is subject to high-frequency homologous recombination and, therefore, is unstable in expression of the tandem repeats. In light of these studies, two concurrent strategies were used to improve immune responses to MUC-1 (Akagi et al. 1997). A recombinant vaccinia virus was constructed encoding a modified miniature MUC-1 gene with only ten tandem repeat sequences to minimize vaccinia-mediated rearrangement (designated rV-MUC-1). An admixture consisting of rV-MUC-1 and rV-B7–1 was then used. The rV-MUC-1 gene product maintained a consistent molecular weight throughout several passages, indicating stability of the inserted gene. Mice inoculated with rV-MUC-1 demonstrated MUC-1-specific cytolytic responses that were further enhanced by admixture with rV-B7–1. In a pulmonary metastases prevention model expressing MUC-1, mice inoculated twice with rV-MUC-1 were protected from establishing metastases. No additive effect on anti-tumor immunity (90% with rV-MUC-1 alone) was observed in mice primed with an admixture of rV-MUC-1 and rV-B7–1 and boosted with rV-MUC-1.

When rV-MUC-1 was used to treat established MUC-1-positive metastases, three administrations of rV-MUC-1 were not sufficient to confer anti-tumor effects (Fig. 4). In contrast, when tumor-bearing mice were primed with an admixture of rV-MUC-1 and rV-B7–1, followed by two boosts with rV-MUC-1, the number of pulmonary metastases decreased significantly (p=.0001), which correlated to 100% survival (Figs. 4, 5). Coexpression of the B7–1 molecule, although not necessary for the induction of an immune response of sufficient magnitude to prevent MUC-1 tumors, was thus essential in a treatment setting.

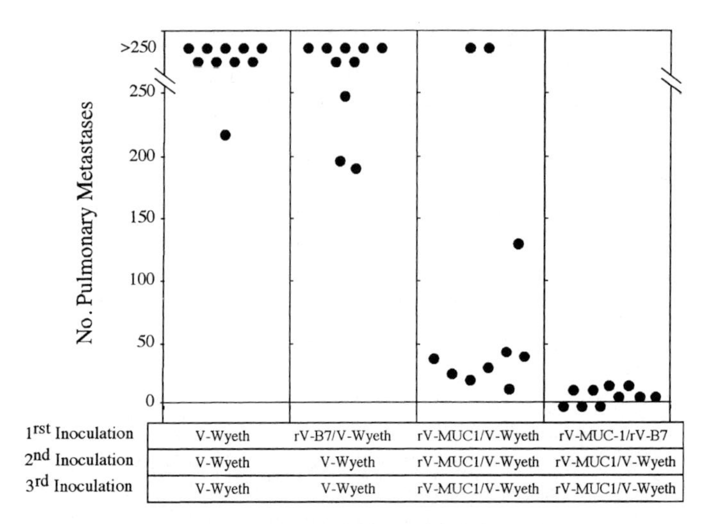

Fig. 4. Treatment of established MUC-1-positive pulmonary metastases by vaccination with an admixture of rV-MUC-1 and rV-B7–1. Groups of 10 mice were transplanted intravenously with 1×10^6 MC38/MUC-1 tumor cells, and tumor cells were allowed to establish for 3 days. Mice were then inoculated every 7 days as indicated. Mice were killed 28 days after tumor transplant, and pulmonary metastatic nodules were stained and enumerated

3.5 Whole-Tumor-Cell Vaccines and Costimulation

Although tumor cells may display TAA on their surface, they usually do not elicit immune responses. One explanation is that, unlike APC, most tumor cells do not naturally express costimulatory molecules and, therefore, are unable to activate the immune system. To address this deficiency, recombinant poxviruses that express a costimulatory molecule can be used to infect tumor cells by direct injection into the tumor itself. Infection results in short-term, or transient, expression of the costimulatory molecules. In preclinical models, such expression has proven sufficient to activate immune responses directed against the antigens naturally present on the surface of tumor cells. Once activated, these immune responses may be capable of recognizing all tumor cells of the same

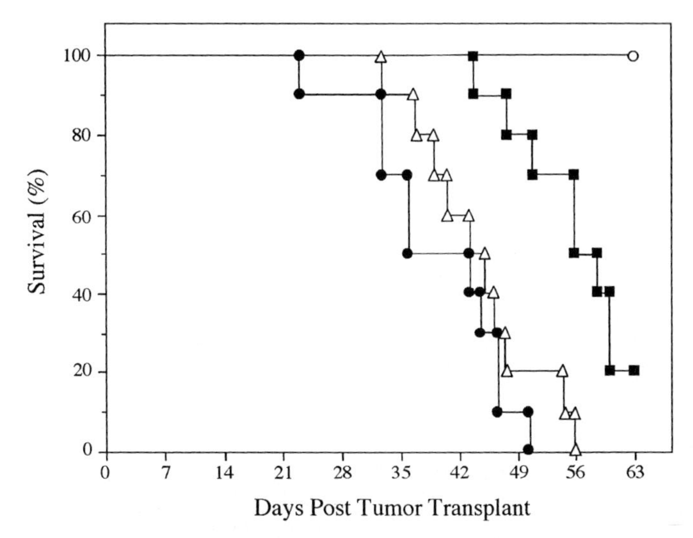

Fig. 5. Survival of mice treated by vaccination with an admixture of rV-MUC-1 and rV-B7–1. Groups of 10 mice were transplanted intravenously with 1×10^6 MC38/MUC-1 tumor cells, and tumors were allowed to establish for 3 days. Mice were inoculated every 7 days as in Fig. 3. Vaccination sequences were: V-WT : V-WT : V-WT (*open triangles*); rV-B7–1/V-WT : V-WT : V-WT (*closed circles*); rV-MUC-1/V-WT : rV-MUC-1/V-WT : rV-MUC-1/V-WT (*closed squares*); and rV-MUC-1/rV-B7–1 : rV-MUC-1/V-WT : rV-MUC-1/V-WT (*open circles*)

type, whether or not these tumor cells also express costimulatory molecules.

Many studies have now revealed the efficacy of whole-tumor-cell vaccines to enhance anti-tumor activity in experimental models (Dranoff 1993; Chen et al. 1994; Hodge et al. 1994; Dunussi-Joannopoulos et al. 1996; Gajewski et al. 1996; Qin and Chatterjee 1996; McLaughlin et al. 1997; Uzendoski et al. 1997; Emtage et al. 1998; Lorenz et al. 1999a,b). Because this approach is not dependent upon the identification of specific tumor antigens, it can be used for the treatment of multiple cancers, including those for which TAAs have not been identified. The advantages of this approach include: (a) efficient delivery of genes to the tumor cells by recombinant poxviruses, (b) ability to phenotypically modify tumor cells in vivo, rather than only after surgical removal, and (c) capability of rapid, efficient ex vivo infection of tumor cells by recombinant poxviruses, after which cells can be X-irradiated and reinjected into the patient.

In some experimental systems, tumor-cell vaccines consisted of live or X-irradiated cells that were highly or moderately immunogenic (Chen et al. 1994). In other studies, tumor cells were shown to be weakly immunogenic or not immunogenic at all (i.e., the tumor cells would grow readily in the host, and vaccines consisting of X-irradiated tumor cells were not capable of inducing anti-tumor immunity). It is in these cases that the insertion and expression of transgenes such as cytokine genes and costimulatory molecule genes may make tumors more immunogenic. The insertion of costimulatory molecule genes is extremely attractive in the case of tumor vaccines because the vast majority of non-hematopoietic tumors do not express T cell costimulatory molecules. To date, most whole-tumor-cell vaccines expressing a transgene via a vector have employed retroviral vectors. Recently, several studies have used poxvirus vectors to infect whole-tumor-cell vaccines. A recent study has compared for the first time the use of a poxvirus vector versus a retroviral vector to express the B7–1 costimulatory molecule transgene in both live and X-irradiated whole-tumor-cell vaccines (Hodge et al. 1996). Both the recombinant retrovirus (R-B7–1) and the recombinant vaccinia (rV-B7–1) induced equivalent expression of B7–1 on the surface of the MC38 murine carcinoma cells slated to be used as vaccine. Wild-type retrovirus (R-WT) and vaccinia virus (V-WT) were used as controls. Using live whole-tumor cells as vaccine, cells transduced via recombinant retrovirus or recombinant vaccinia virus expressing B7–1 equally induced protection against challenge with native MC38 tumor cells. Upon rechallenge with native tumor cells 40 days later, however, the R-B7–1 vaccine was shown to be less effective than the rV-B7–1 whole-tumor-cell vaccine. These experiments were also conducted using X-irradiated tumor cells as vaccine. Again, the rV-B7–1 vector-infected X-irradiated tumor cells were superior to the vaccine prepared with R-B7–1. Comparative studies have also been conducted in which X-irradiated tumor-cell vaccines were administered to mice that developed experimental lung metastases. In these therapy studies, all mice receiving X-irradiated native tumor cells developed more than 200 metastatic nodules in the lung, similar to the result seen in non-vaccinated mice. All mice receiving vaccine consisting of

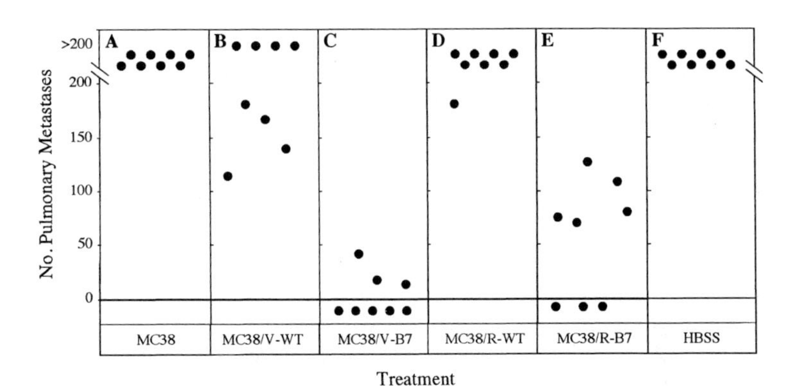

Fig. 6A–F. Treatment of lung metastases employing X-irradiated whole-tumor-cell vaccines in mice that had been administered wild-type vaccinia (V-WT). Fourteen days prior to tumor burden, all mice received 1×10^7 pfu V-WT. Groups of eight mice were then inoculated intravenously with 1×10^6 native MC38 cells on day 0. On days 3, 10, and 17, mice received as vaccine X-irradiated MC38 tumor cells (**A**), X-irradiated tumor cells previously infected with V-WT (**B**), V-B7–1 (**C**), R-WT (**D**), or R-B7–1 (**E**), or no tumor cells (HBSS buffer; **F**). On day 28, mice were killed and lung metastases were counted

X-irradiated tumor cells infected with V-WT or R-WT also developed lung metastases. Mice receiving the X-irradiated rV-B7–1 tumor-cell vaccine experienced a statistically significant reduction in lung metastases compared with those mice receiving vaccine infected with R-WT. Similarly, mice receiving the X-irradiated rV-B7–1 tumor vaccine had a statistically significant reduction in the development of lung metastases compared with those mice receiving X-irradiated V-WT vaccine. And mice receiving the X-irradiated rV-B7–1 vaccine experienced a statistically significant reduction in lung metastases compared with mice receiving the X-irradiated vaccine infected with R-B7–1.

One of the concerns inherent in using a recombinant vaccinia-based vaccine is that previous immunity, for example via the smallpox vaccine or prior poxvirus vaccinations, would inhibit its effectiveness (Demkowicz et al. 1996; Hodge et al. 1997). To determine if prior exposure to vaccinia would inhibit the anti-tumor efficacy of X-irradiated whole-tumor-cell vaccines with the rV-B7–1 vector, the studies described above were performed in mice that had received 10^7 pfu V-WT 17 days prior to vaccination. This dose and interval had been shown to lead to the development of substantial anti-vaccinia immune responses. As seen in Fig. 6, the X-irradiated R-B7–1 tumor-cell vaccine remained statistically significant in its therapeutic effectiveness compared with the retroviral vaccine. Indeed, in these types of whole-tumor-cell vaccines, there is no need for virus replication and, thus, there should be no inhibition of anti-tumor efficacy by anti-poxvirus immune responses. Moreover, such immune responses may actually enhance the anti-tumor effect in this setting. This phenomenon formed the basis for the use of oncolysates, employing extracts of vaccinia as vaccines (Wallack et al. 1998). As previously pointed out, such preparations may have been less than optimal because no costimulatory molecules were present to provide the required second signal for enhanced T cell activation. Studies have been reported in which anti-CTLA-4 monoclonal antibodies (MAb) were used to inhibit growth of tumors in some murine models but not in others, particularly those tumors that are poorly immunogenic (Leach et al. 1996; Yang et al. 1997; Greenfield et al. 1998). The use of anti-CTLA-4 MAb was shown to be ineffectual in the tumor model used in these studies (Mokyr et al. 1998).

Pros and cons exist in the use of any vector for anti-cancer vaccine applications. Advantages of using a retroviral vector in whole-tumor-cell vaccines include stable integration, transgene expression by all cells if the cells are drug-selected and cloned, and a proposed lack of immunogenicity. Most studies have used murine retroviruses in murine systems. It is not clear at this point if anti-retroviral immunity in humans develops after administration of tumor-cell vaccines containing retroviruses. There are several potential disadvantages in using a retroviral vector that do not exist when one is employing poxvirus vectors in whole-tumor-cell vaccines. Poxviruses do not require a lengthy drug-selection or cloning process, since it has been shown that they can express the transgene efficiently in greater than 95% of cells within 5 h (Hodge et al. 1994). Moreover, unlike retroviruses, poxviruses do not require cell division to express transgenes. While the use of retrovirus vectors is very efficient for established tumor cell lines that rapidly divide in vitro and in vivo, this is not the case for cells derived from human tumor biopsies or for human tumors in situ. For example, it is extremely difficult to propagate tumor cells from biopsies in cases of breast and colon carcinoma, and cells do not divide rapidly in situ. Perhaps the main advantage in using poxviruses in whole-tumor-cell vaccines is the fact that one can insert multiple genes into a poxvirus vector, which is not possible for most other types of vectors. Vectors containing three or more costimulatory molecules possess great potential for this approach and clearly merit further investigation.

3.6 Construction and Characterization of a Recombinant Vaccinia Virus Expressing ICAM-1: Induction and Potentiation of Anti-Tumor Responses

ICAM-1 has been associated with cellular migration into inflammatory sites and with facilitating interactions between lymphocytes and tumor targets in the pathway of cell-mediated cytotoxicity. More recently, ICAM-1 (murine) has been implicated frequently in the costimulation of T cell functions such as antigen-dependent T cell proliferation. Previous murine studies have shown that the introduction of the ICAM-1 gene into tumor cells using retroviral vectors led to enhanced anti-tumor responses. Studies have now shown the construction, characterization and immunological consequences of a recombinant vaccinia virus expressing murine ICAM-1 (rV-ICAM-1; Uzendoski et al. 1997). The infection of tumor cells with rV-ICAM-1 resulted in the expression of functional ICAM-1. Infected tumors provide accessory or secondary signals to lymphoblasts in vitro, resulting in enhanced cytokine production or alloreactive cytotoxic T lymphocyte (CTL) activity. In vivo, it was demonstrated that weakly immunogenic syngeneic tumors, infected with and expressing rV-ICAM-1, were rejected by immunocompetent hosts. Furthermore, vaccination with rV-ICAM-1-infected tumors resulted in the rejection of subsequent tumor challenge, providing evidence for recall response and immunological memory (Fig. 7). These studies indicated the utility of a recombinant vaccinia virus to deliver and efficiently express ICAM-1 molecules on tumor cells for potential gene therapy and recombinant approaches to cancer immunotherapy (Uzendoski et al. 1997).

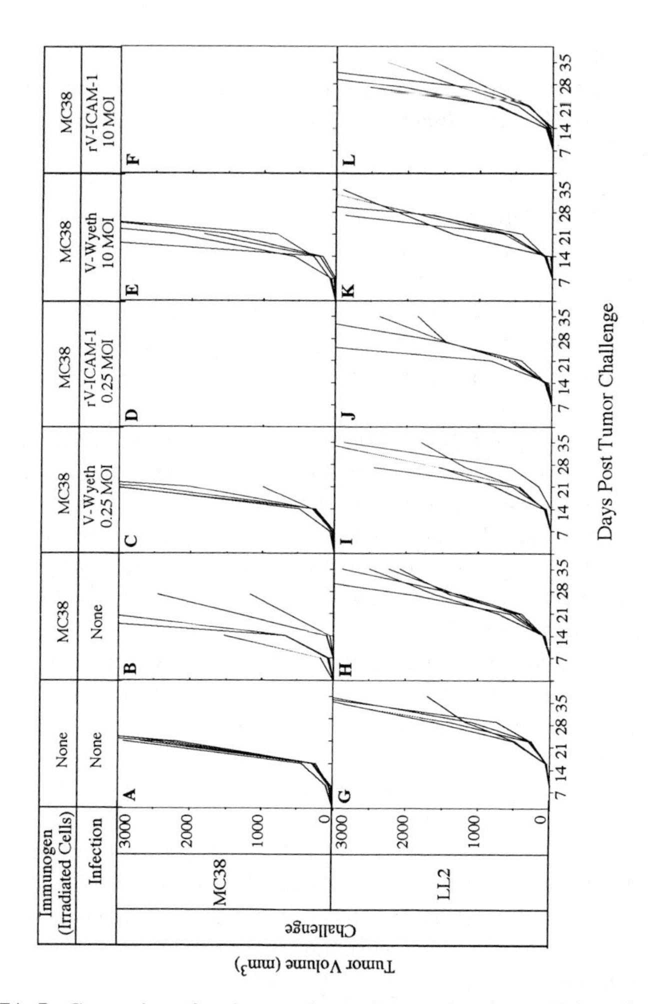

Fig. 7A–L. Generation of anti-tumor immunity requires intercellular adhesion molecule-1 (ICAM-1) to be expressed by the primary tumor immunogen. C57BL/6 mice were injected on the left flank with 3×10^5 γ-irradiated MC38 tumor cells as shown under conditions described (Dunussi-Joannopoulos et al. 1996). By day 21, when primary tumors were not palpable in any of these mice, they were challenged with 3×10^5 viable MC38 or LL2 control tumor cells

3.7 Induction of Anti-Tumor Immunity Elicited by Tumor Cells Expressing a Murine LFA-3 Analog via a Recombinant Vaccinia Virus

Murine LFA-3 (CD48) is the structural and functional analog of human LFA-3 (CD58; Selvaraj et al. 1987; Wong et al. 1990) and is a member of the CD2 family within the immunoglobulin gene superfamily. It is widely expressed on hematopoietic cells including thymocytes, splenocytes, B cells, T cells, and Langerhans' cells but not on fibroblasts and other non-hematopoietic cells (Wong et al. 1990). The natural ligand of LFA-3 is CD2, which is expressed on thymocytes, T cells, B cells, and natural killer (NK) cells in the mouse (Meuer et al. 1984; Yagita et al. 1988; Altevogt et al. 1989; Guckel et al. 1991; Davis and van der Merwe 1996). CD2 engagement is also involved in T cell development, inhibition of B cell and T cell apoptosis, CTL function, and regulation of cytolytic activities of NK cells (Anasetti et al. 1987; van de Griend et al. 1987; Chavin et al. 1993; Genaro et al. 1994; Ayroldi et al. 1997). These features make LFA-3 an attractive molecule to exploit for induction of anti-tumor immunity.

To determine the effect of LFA-3 expression on the immunogenicity of tumor cells, a recombinant vaccinia virus containing the murine LFA-3 gene (designated rV-LFA-3) was constructed (Lorenz et al. 1999a). rV-LFA-3 was shown to be functional in vitro in terms of expression of LFA-3, T cell proliferation, adhesion, and cytotoxicity. Subcutaneous inoculation of rV-LFA-3-infected murine colon adenocarcinoma tumor cells (MC38) into immunocompetent syngeneic C57BL/6 mice resulted in complete lack of tumor growth (Lorenz et al. 1999a). Inoculation of MC38 cells infected with equal doses of control wild-type vaccinia virus resulted in tumor growth in all animals. In addition, partial immunological protection was demonstrated against subsequent challenge with uninfected parental tumor cells up to 56 days after vaccination with rV-LFA-3-infected cells. Anti-tumor memory was also demonstrated by using γ-irradiated MC38 cells and cells from another carcinoma model (CT26). These studies show that expression of LFA-3 via a poxvirus vector can be used to induce anti-tumor immunity (Table 2; Lorenz et al. 1999a).

Table 2. Effect of leukocyte function-associated antigen-3 (LFA-3) expression on CT26 tumor-cell immunogenicity[a]

Immunogen	Mice/group	Tumor volume (mm^3±SEM)	*P* value[b]
HBSS	5	6864±1269	0.083[c]
V-WT-infected tumor cells (γ-irradiation)	5	2255±376	0.083[d]
rV-LFA-3-infected tumor cells (γ-irradiation)	8	373±1200	0.0001[c] <0 .0001[d]

[a]C57BL/6 mice were vaccinated 3 times subcutaneously at 14-day intervals with γ-irradiated MC38 cells infected with a multiplicity of infection of 10 of the indicated virus, then challenged with parental MC38 cells (5×10^5/mouse) 4 weeks later. Tumors were measured 21 days after tumor challenge
[b]*P* values were calculated at 95% by ANOVA
[c]Compared with the V-WT group
[d]Compared with HBSS group

3.8 The Next Generation: a Vector Containing a Triad of Costimulatory Molecules

As mentioned above, one of the major advantages of the use of recombinant poxvirus vaccines is the ability to insert multiple transgenes. Using retroviral vector infection of tumor cells and multiple drug selections, it has been shown that the insertion of two costimulatory molecule genes into tumors can produce additive or synergistic activation of T cells (Cavallo et al. 1995; Wingren et al. 1995; Li et al. 1996; Parra et al. 1997). Newly designed and developed poxvirus constructs have been shown to be capable of expressing a triad of costimulatory molecules (B7–1, ICAM-1, and LFA-3, designated TRICOM; Hodge et al. 1999a). Within 5 h of infection, tumor cells infected with either recombinant fowlpox (rF)-TRICOM or rV-TRICOM were shown to express all three costimulatory molecules on the cell surface. Using Con A as a generic signal 1, a panel of tumor cells was created that expressed each of the three costimulatory molecules alone and together (as the TRICOM construct) to provide costimulatory signals. Both $CD4^+$ and $CD8^+$ T cells were isolated, and their ability to be stimulated was analyzed. The stratification of stimulator cell effects of the costimulatory molecule(s) on proliferation was similar for both $CD4^+$ and $CD8^+$ T cells (Fig. 8). As can be seen, the TRICOM vector provided the most potent stimulation of both $CD4^+$ and $CD8^+$ T cells. Moreover, these effects were clearly synergistic, not additive. Cytokine expression from $CD4^+$ and $CD8^+$ T cells stimulated with single or multiple costimulatory molecules was also analyzed at the RNA level utilizing the multiprobe RNase protection assay. A representative radiographic profile and quantitative analysis from two independent experiments are depicted (Fig. 9). IL-2 and IFN-γ expression levels were highest in $CD4^+$ T cells stimulated with TRICOM when compared with $CD4^+$ cells stimulated with MC38 cells expressing any single costimulatory molecule (Fig. 9B). Expression of cytokine genes was also analyzed in stimulated $CD8^+$ T cells. Of the cytokine RNAs analyzed, IL-2 and particularly IFN-γ levels were significantly higher when these cells were stimulated with TRICOM, compared with T cells stimulated with MC38 cells expressing any single costimulatory molecule (Fig. 9C). Thus, the predominant synergistic effect of the TRICOM vector in cytokine production was IL-2 in $CD4^+$ cells and IFN-γ in $CD8^+$ T cells.

These studies thus indicated that poxviruses containing as many as three costimulatory molecules as transgenes can rapidly and efficiently activate T cell populations to levels far greater than those achieved when any one or two of these costimulatory molecules is used. Previous toxicology studies analyzing the effects of multiple administrations of rV-B7–1 in mice revealed no toxicity, including no evidence of autoimmunity (Freund et al. 2000). Similar studies are ongoing using recombinant TRICOM poxvirus vectors. The ability to achieve this new threshold of T cell activation using vectors containing multiple costimulatory molecules obviously has broad implications in the design and development of anti-cancer vaccines since there is overwhelming evidence that the vast majority of TAAs are weakly immunogenic.

Two clinical trials employing ALVAC-CEA/B7–1 as a vaccine in patients with advanced cancer have recently been completed (Lee at el. 1999; von Mehren et al. 1999). These were the first clinical trials to use a recombinant poxvirus vector containing a costimulatory molecule gene and a TAA gene. Using the ELISPOT assay for IFN-γ production and analyzing peripheral blood mononuclear cells (PBMC) after less than 24 h in culture, substantial increases in CEA-specific T cells were observed postvaccination compared to prevaccination. As a control, no

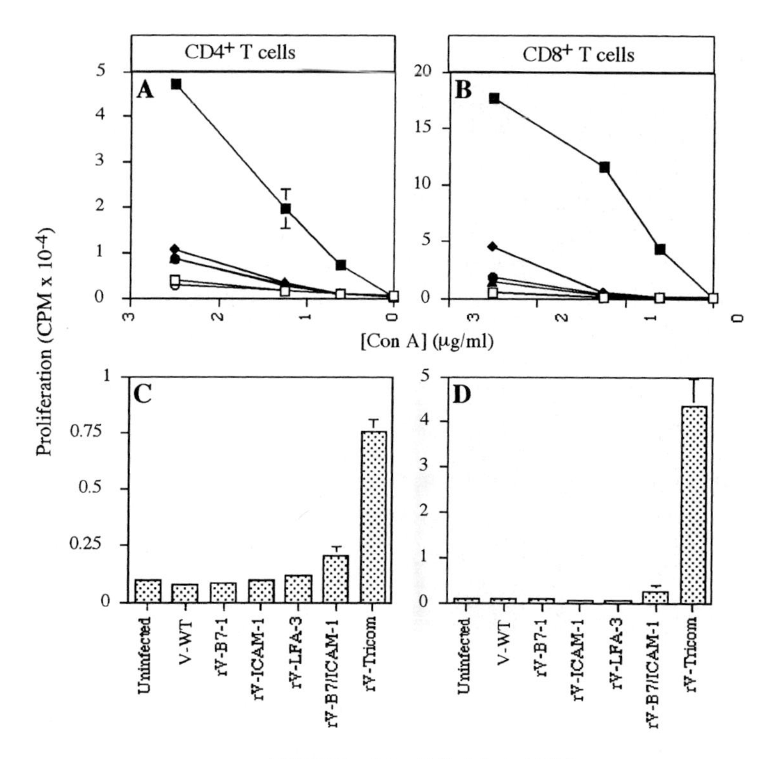

Fig. 8A–D. Effect of costimulation on specific T cell populations. Murine CD4^{+} (**A**) or CD8^{+} T cells (**B**) were cocultured with uninfected MC38 cells (*open circle*) or cells infected with V-WT (*open squares*), rV-LFA-3 (*closed triangles*), rV-ICAM-1 (*closed circles*), rV-B7–1 (*closed diamonds*), or rV-TRICOM (*closed squares*) at a 10:1 ratio for 48 h in the presence of various concentrations of Con A. **C,D** Proliferative responses of purified CD4^{+} and CD8^{+} cells, respectively, when cocultured in the presence of vector-infected MC38 stimulator cells at a low Con A concentration (0.625 μg/ml)

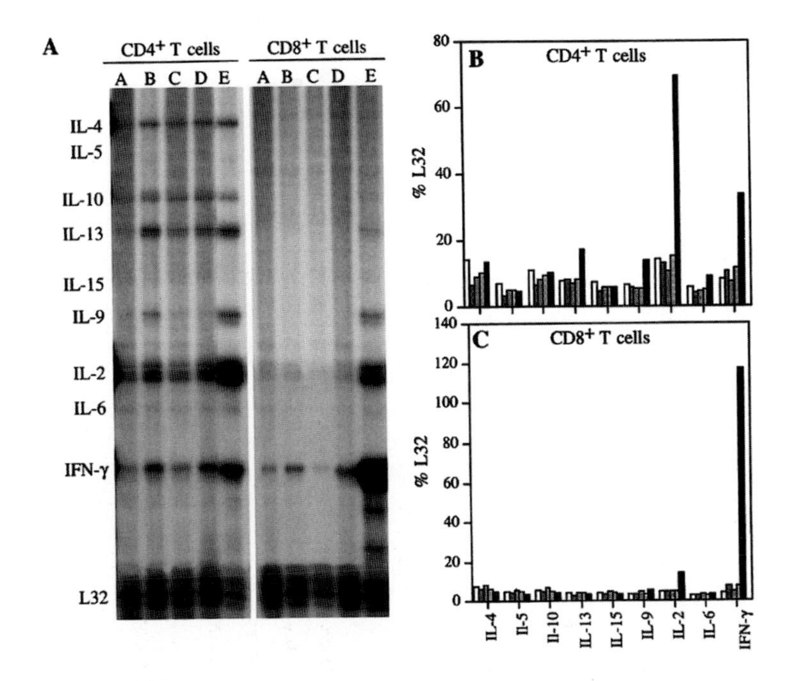

Fig. 9A–C. Effect of costimulation on cytokine RNA expression. **A** Murine $CD4^+$ or $CD8^+$ T cells were cocultured with MC38 stimulator cells infected with V-Wyeth (*A*), rV-B7–1(*B*), rV-ICAM-1 (*C*), rV-LFA-3 (*D*), or rV-B7–1/ICAM-1/LFA-3 (*E*) at a T cell to stimulator cell ratio of 10:1 for 24 h in the presence of 2.5 µg/ml Con A. Following culture, T cell RNA was analyzed by multiprobe RNase protection assay. The quantitative representation of results from the autoradiograph is normalized for expression of the housekeeping gene L32 in **B** ($CD4^+$ cells) and **C** ($CD8^+$ cells). Order of *histogram bars* (from *left* to *right*) is MC38/V-Wyeth, MC38/B7–1, MC38/ICAM-1, MC38/LFA-3, and MC38/B7–1/ICAM-1/LFA-3

changes in T cell responses were observed to a Flu peptide from PBMC obtained pre- and postimmunization.

Perhaps the most intriguing advantage of using recombinant poxvirus vectors in anti-cancer vaccines is the ability to insert multiple transgenes; this can include the use of multiple TAA genes, cytokine genes, and/or T cell costimulatory molecule genes. The belief that the vast majority of TAA genes are weak immunogens supports this approach. Underscoring these vectors' significant potential as vaccines is the demonstration that recombinant poxviruses can efficiently express three T cell costimulatory molecules in the same APC and activate T cells to levels not previously achievable with the use of any one or two of these molecules. Modes of application of pox-vector-based vaccines potentially include direct administration into patients as a classic vaccine, infection of whole-tumor cells either in vitro or in situ, or the in vitro infection of APC such as dendritic cells to enhance the effectiveness of this approach.

References

Akagi J, Hodge JW, McLaughlin JP, Gritz L, Mazzara G, Kufe D, Schlom J, Kantor JA (1997) Therapeutic antitumor response after immunization with an admixture of recombinant vaccinia viruses expressing a modified MUC1 gene and the murine T-cell costimulatory molecule B7. J Immunother 20:38–47

Altevogt P, Michaelis M, Kyewski B (1989) Identical forms of the CD2 antigen expressed by mouse T and B lymphocytes. Eur J Immunol 19:1509–1512

Anasetti C, Martin PJ, June CH, Hellstrom KE, Ledbetter JA, Rabinovitch PS, Morishita Y, Hellstrom I, Hansen JA (1987) Induction of calcium flux and enhancement of cytolytic activity in natural killer cells by cross-linking of the sheep erythrocyte binding protein (CD2) and the Fc-receptor (CD16). J Immunol 139:1772–1779

Ayroldi E, Migliorati G, Cannarile L, Moraca R, Delfino DV, Riccardi C (1997) CD2 rescues T cells from T-cell receptor/CD3 apoptosis: a role for the Fas/Fas-L system. Blood 89:3717–3726

Bei R, Kantor J, Kashmiri SV, Abrams S, Schlom J (1994) Enhanced immune responses and anti-tumor activity by baculovirus recombinant carcinoembryonic antigen (CEA) in mice primed with the recombinant vaccinia CEA. J Immunother Emphasis Tumor Immunol 16:275–282

Belshe RB, Gorse GJ, Mulligan MJ, Evans TG, Keefer MC, Excler JL, Duliege AM, Tartaglia J, Cox WI, McNamara J, Hwang KL, Bradney A, Montefiori D, Weinhold KJ (1998) Induction of immune responses to HIV-1 by canarypox virus (ALVAC) HIV-1 and gp120 SF-2 recombinant vaccines in uninfected volunteers. NIAID AIDS Vaccine Evaluation Group. AIDS 12:2407–2415

Carroll MW, Overwijk WW, Chamberlain RS, Rosenberg SA, Moss B, Restifo NP (1997) Highly attenuated modified vaccinia virus Ankara (MVA) as an effective recombinant vector: a murine tumor model. Vaccine 15:387–394

Cavallo F, Martin-Fontecha A, Bellone M, Heltai S, Gatti E, Tornaghi P, Freschi M, Forni G, Dellabona P, Casorati G (1995) Co-expression of B7–1 and ICAM-1 on tumors is required for rejection and the establishment of a memory response. Eur J Immunol 25:1154–1162

Chavin KD, Qin L, Lin J, Kaplan AJ, Bromberg JS (1993) Anti-CD2 and anti-CD3 monoclonal antibodies synergize to prolong allograft survival with decreased side effects. Transplantation 55:901–908

Chen L, McGowan P, Ashe S, Johnston J, Li Y, Hellstrom I, Hellstrom KE (1994) Tumor immunogenicity determines the effect of B7 costimulation on T cell-mediated tumor immunity. J Exp Med 179:523–532

Clements-Mann ML, Weinhold K, Matthews TJ, Graham BS, Gorse GJ, Keefer MC, McElrath MJ, Hsieh RH, Mestecky J, Zolla-Pazner S, et al (1998) Immune responses to human immunodeficiency virus (HIV) type 1 induced by canarypox expressing HIV-1MN gp120, HIV-1SF2 recombinant gp120, or both vaccines in seronegative adults. NIAID AIDS Vaccine Evaluation Group. J Infect Dis 177:1230–1246

Cole DJ, Wilson MC, Baron PL, O'Brien P, Reed C, Tsang KY, Schlom J (1996) Phase I study of recombinant CEA vaccinia virus vaccine with post vaccination CEA peptide challenge. Hum Gene Ther 7:1381–1394

Davis SJ, Merwe PA van der (1996) The structure and ligand interactions of CD2: implications for T-cell function. Immunol Today 17:177–187

Demkowicz WE Jr, Littaua RA, Wang J, Ennis FA (1996) Human cytotoxic T-cell memory: long-lived responses to vaccinia virus. J Virol 70:2627–2631

Dranoff G, Jaffee E, Lazenby A, Golumbek P, Levitsky H, Brose K, Jackson V, Hamada H, Pardoll D, Mulligan RC (1993) Vaccination with irradiated tumor cells engineered to secrete murine granulocyte-macrophage colony-stimulating factor stimulates potent, specific, and long-lasting anti-tumor immunity. Proc Natl Acad Sci USA 90:3539–3543

Dunussi-Joannopoulos K, Weinstein HJ, Nickerson PW, Strom TB, Burakoff SJ, Croop JM, Arceci RJ (1996) Irradiated B7–1 transduced primary acute myelogenous leukemia (AML) cells can be used as therapeutic vaccines in murine AML. Blood 87:2938–2946

Egan MA, Pavlat WA, Tartaglia J, Paoletti E, Weinhold KJ, Clements ML, Siliciano RF (1995) Induction of human immunodeficiency virus type 1 (HIV-1)-specific cytolytic T lymphocyte responses in seronegative adults by a nonreplicating, host-range-restricted canarypox vector (ALVAC) carrying the HIV-1MN env gene. J Infect Dis 171:1623–1627

Emtage PC, Wan Y, Muller W, Graham FL, Gauldie J (1998) Enhanced interleukin-2 gene transfer immunotherapy of breast cancer by coexpression of B7–1 and B7–2. J Interferon Cytokine Res 18:927–937

Freund YR, Mirsalis JC, Fairchild DG, Brune J, Hokama LA, Schindler-Horvat J, Tomaszewski JE, Hodge JW, Schlom J, Kantor JA, Tyson CA, Donohue SJ (2000) Vaccination with a recombinant vaccinia vaccine containing the B7–1 costimulatory molecule causes no significant toxicity and enhances T cell mediated cytotoxicity. Int J Cancer 85:508–517

Fries LF, Tartaglia J, Taylor J, Kauffman EK, Meignier B, Paoletti E, Plotkin S (1996) Human safety and immunogenicity of a canarypox-rabies glycoprotein recombinant vaccine: an alternative poxvirus vector system. Vaccine 14:428–434

Gajewski TF, Fallarino F, Uyttenhove C, Boon T (1996) Tumor rejection requires a CTLA4 ligand provided by the host or expressed on the tumor: superiority of B7–1 over B7–2 for active tumor immunization. J Immunol 156:2909–2917

Genaro AM, Gonzalo JA, Bosca L, Martinez C (1994) CD2-CD48 interaction prevents apoptosis in murine B lymphocytes by up-regulating bcl-2 expression. Eur J Immunol 24:2515–2521

Graham BS, Belshe RB, Clements ML, Dolin R, Corey L, Wright PF, Gorse GJ, Midthun K, Keefer MC, Roberts NJ Jr, et al (1992) Vaccination of vaccinia-naive adults with human immunodeficiency virus type 1 gp160 recombinant vaccinia virus in a blinded, controlled, randomized clinical trial. The AIDS Vaccine Clinical Trials Network. J Infect Dis 166:244–252

Graham BS, Gorse GJ, Schwartz DH, Keefer MC, McElrath MJ, Matthews TJ, Wright PF, Belshe RB, Clements ML, Dolin R, et al (1994) Determinants of antibody response after recombinant gp160 boosting in vaccinia-naive volunteers primed with gp160-recombinant vaccinia virus. The National Institute of Allergy and Infectious Diseases AIDS Vaccine Clinical Trials Network. J Infect Dis 170:782–786

Greenfield EA, Nguyen KA, Kuchroo VK (1998) CD28/B7 costimulation: a review. Crit Rev Immunol 18:389–418

Griend RJ van de, Bolhuis RL, Stoter G, Roozemond RC (1987) Regulation of cytolytic activity in CD3– and CD3+ killer cell clones by monoclonal antibodies (anti-CD16, anti-CD2, anti-CD3) depends on subclass specificity of target cell IgG-FcR. J Immunol 138:3137–3144

Guckel B, Berek C, Lutz M, Altevogt P, Schirrmacher V, Kyewski BA (1991) Anti-CD2 antibodies induce T cell unresponsiveness in vivo. J Exp Med 174:957–967
Hodge JW, Abrams S, Schlom J, Kantor JA (1994) Induction of antitumor immunity by recombinant vaccinia viruses expressing B7–1 or B7–2 costimulatory molecules. Cancer Res 54:5552–5555
Hodge JW, McLaughlin JP, Abrams SI, Shupert WL, Schlom J, Kantor JA (1995) Admixture of a recombinant vaccinia virus containing the gene for the costimulatory molecule B7 and a recombinant vaccinia virus containing a tumor-associated antigen gene results in enhanced specific T-cell responses and antitumor immunity. Cancer Res 55:3598–3603
Hodge JW, McLaughlin JP, Kantor JA, Schlom J (1997) Diversified prime and boost protocols using recombinant vaccinia virus and recombinant non-replicating avian pox virus to enhance T-cell immunity and antitumor responses. Vaccine 15:759–768
Hodge JW, Sabzevari H, Lorenz MGO, Yafal AG, Gritz L, Schlom J (1999a) A triad of costimulatory molecules synergize to amplify T-cell activation. Cancer Res 59:5800–5807
Hodge JW, Schlom J (1999b) Comparative studies of a retrovirus versus a poxvirus vector in whole-tumor-cell vaccines. Cancer Res 59:5106–5111
Kahn M, Sugawara H, McGowan P, Okuno K, Nagoya S, Hellstrom KE, Hellstrom I, Greenberg P (1991) CD4+ T cell clones specific for the human p97 melanoma-associated antigen can eradicate pulmonary metastases from a murine tumor expressing the p97 antigen. J Immunol 146:3235–3241
Kalus RM, Kantor JA, Gritz L, Gomez Yafal A, Mazzara GP, Schlom J, Hodge JW (1999) The use of combination vaccinia vaccines and dual-gene vaccinia vaccines to enhance antigen-specific T-cell immunity via T-cell costimulation. Vaccine 17:893–903
Leach DR, Krummel MF, Allison JP (1996) Enhancement of antitumor immunity by CTLA-4 blockade. Science 271:1734–1736
Lee DS, Conkright W, Horig HE (1999) Preliminary results of ALVAC-CEA-B7.1 phase I vaccine trial in patients with metastatic CEA-expressing tumors. Am Soc Clin Oncol (ASCO) Meeting, Office of J Clin Oncol, Chestnut Hill, Mass., USA
Li Y, Hellstrom KE, Newby SA, Chen L (1996) Costimulation by CD48 and B7–1 induces immunity against poorly immunogenic tumors. J Exp Med 183:639–644
Lorenz MG, Kantor JA, Schlom J, Hodge JW (1999a) Induction of anti-tumor immunity elicited by tumor cells expressing a murine LFA-3 analog via a recombinant vaccinia virus. Hum Gene Ther 10:623–631

Lorenz MG, Kantor JA, Schlom J, Hodge JW (1999b) Antitumor immunity elicited by a recombinant vaccinia virus expressing CD70 (CD27L). Hum Gene Ther 10:1095–1103

Marshall JL, Richmond E, Pedicano J (1999) Phase I/II trial of vaccinia-CEA and ALVAC-CEA in patients with advanced CEA-bearing tumors. Am Soc Clin Oncol (ASCO) Meeting

McAneny D, Ryan CA, Beazley RM, Kaufman HL (1996) Results of a phase I trial of a recombinant vaccinia virus that expresses carcinoembryonic antigen in patients with advanced colorectal cancer. Ann Surg Oncol 3:495–500

McLaughlin JP, Abrams S, Kantor J, Dobrzanski MJ, Greenbaum J, Schlom J, Greiner JW (1997) Immunization with a syngeneic tumor infected with recombinant vaccinia virus expressing granulocyte-macrophage colony-stimulating factor (GM-CSF) induces tumor regression and long-lasting systemic immunity. J Immunother 20:449–459

Mehren M von, Davies M, Davies VR (1999) Phase I trial with ALVAC-CEA B7.1 immunization in advanced CEA-expressing adenocarcinomas. Am Soc Clin Oncol (ASCO) Meeting

Meuer SC, Hussey RE, Fabbi M, Fox D, Acuto O, Fitzgerald KA, Hodgdon JC, Protentis JP, Schlossman SF, Reinherz EL (1984) An alternative pathway of T-cell activation: a functional role for the 50 kd T11 sheep erythrocyte receptor protein. Cell 36:897–906

Mokyr MB, Kalinichenko T, Gorelik L, Bluestone JA (1998) Realization of the therapeutic potential of CTLA-4 blockade in low-dose chemotherapy-treated tumor-bearing mice. Cancer Res 58:5301–5304

Montefiori DC, Graham BS, Kliks S, Wright PF (1992) Serum antibodies to HIV-1 in recombinant vaccinia virus recipients boosted with purified recombinant gp160. NIAID AIDS Vaccine Clinical Trials Network. J Clin Immunol 12:429–439

Moss B (1996) Genetically engineered poxviruses for recombinant gene expression, vaccination, and safety. Proc Natl Acad Sci USA 93:11341–11348

Moss B, Carroll MW, Wyatt LS, Bennink JR, Hirsch VM, Goldstein S, Elkins WR, Fuerst TR, Lifson JD, Piatak M, Restifo NP, Overwijk W, Chamberlain R, Rosenberg SA, Sutter G (1996) Host range restricted, non-replicating vaccinia virus vectors as vaccine candidates. Adv Exp Med Biol 397:7–13

Ockenhouse CF, Sun PF, Lanar DE, Wellde BT, Hall BT, Kester K, Stoute JA, Magill A, Krzych U, Farley L et al (1998) Phase I/IIa safety, immunogenicity, and efficacy trial of NYVAC-Pf7, a pox-vectored, multiantigen, multistage vaccine candidate for *Plasmodium falciparum* malaria. J Infect Dis 177:1664–1673

Parra E, Wingren AG, Hedlund G, Kalland T, Dohlsten M (1997) The role of B7–1 and LFA-3 in costimulation of CD8+ T cells. J Immunol 158:637–642

Qin H, Chatterjee SK (1996) Cancer gene therapy using tumor cells infected with recombinant vaccinia virus expressing GM-CSF. Hum Gene Ther 7:1853–1860

Sanda MG, Smith DC, Charles LG, Hwang C, Pienta KJ, Schlom J, Milenic D, Panicali D, Montie JE (1999) Recombinant vaccinia-PSA (PROSTVAC) can induce a prostate-specific immune response in androgen-modulated human prostate cancer. Urology 53:260–266

Selvaraj P, Plunkett ML, Dustin M, Sanders ME, Shaw S, Springer TA (1987) The T lymphocyte glycoprotein CD2 binds the cell surface ligand LFA-3. Nature 326:400–403

Stienlauf S, Shoresh M, Solomon A, Lublin-Tennenbaum T, Atsmon Y, Meirovich Y, Katz E (1999) Kinetics of formation of neutralizing antibodies against vaccinia virus following re-vaccination. Vaccine 17:201–204

Tartaglia J, Excler JL, El Habib R, Limbach K, Meignier B, Plotkin S, Klein M (1998) Canarypox virus-based vaccines: prime-boost strategies to induce cell-mediated and humoral immunity against HIV. AIDS Res Hum Retroviruses 14 (suppl) 3:S291–S298

Tubiana R, Gomard E, Fleury H, Gougeon ML, Mouthon B, Picolet H, Katlama C (1997) Vaccine therapy in early HIV-1 infection using a recombinant canarypox virus expressing gp160MN (ALVAC-HIV): a double-blind controlled randomized study of safety and immunogenicity. AIDS 11:819–820

Uzendoski K, Kantor JA, Abrams SI, Schlom J, Hodge JW (1997) Construction and characterization of a recombinant vaccinia virus expressing murine intercellular adhesion molecule-1: induction and potentiation of antitumor responses. Hum Gene Ther 8:851–860

Wallack MK, Sivanandham M, Balch CM, Urist MM, Bland KI, Murray D, Robinson WA, Flaherty L, Richards JM, Bartolucci AA, Rosen L (1998) Surgical adjuvant active specific immunotherapy for patients with stage III melanoma: the final analysis of data from a phase III, randomized, double-blind, multicenter vaccinia melanoma oncolysate trial. J Am Coll Surg 187:69–77

Wingren AG, Parra E, Varga M, Kalland T, Sjogren HO, Hedlund G, Dohlsten M (1995) T cell activation pathways: B7, LFA-3, and ICAM-1 shape unique T cell profiles. Crit Rev Immunol 15:235–253

Wong YW, Williams AF, Kingsmore SF, Seldin MF (1990) Structure, expression, and genetic linkage of the mouse BCM1 (OX45 or Blast-1) antigen. Evidence for genetic duplication giving rise to the BCM1 region on mouse chromosome 1 and the CD2/LFA3 region on mouse chromosome 3. J Exp Med 171:2115–2130

Yagita H, Okumura K, Nakauchi H (1988) Molecular cloning of the murine homologue of CD2. Homology of the molecule to its human counterpart T11. J Immunol 140:1321–1326

Yang YF, Zou JP, Mu J, Wijesuriya R, Ono S, Walunas T, Bluestone J, Fujiwara H, Hamaoka T (1997) Enhanced induction of antitumor T-cell responses by cytotoxic T lymphocyte-associated molecule-4 blockade: the effect is manifested only at the restricted tumor-bearing stages. Cancer Res 57:4036–4041

4 Application of T Cell Immunotherapy for Human Viral and Malignant Diseases

S. R. Riddell, E. H. Warren, D. Lewinsohn, H. Mutimer, M. Topp, L. Cooper, R. de Fries, P. D. Greenberg

4.1 Introduction

Improvements in our understanding of the molecular basis for T cell recognition of virus-infected cells and tumors, and of the signals involved in eliciting and maintaining a competent immune response has led to new efforts to bolster host T cell immunity in settings where deficient responses permit disease progression. The identification of viral antigens and antigens expressed by tumors has led to efforts to develop adoptive immunotherapy with T cell clones as a therapeutic approach to restore or augment host responses (Riddell and Greenberg 1995). The early results of clinical studies of T cell therapy for viral diseases have been encouraging and this approach is now being developed for the treatment of patients with leukemia that recurs after allogeneic bone marrow transplant (BMT) and patients with solid tumors.

4.2 Immunotherapy of Viral Diseases

4.2.1 Cytomegalovirus (CMV) Disease in Allogeneic BMT Recipients

CMV is a herpes virus that infects the majority of the population during childhood or adolescence and then persists in the host for life. In individuals with a normal immune system, persistent infection is controlled by the development and maintenance of host virus-specific immune responses. Patients with iatrogenic or acquired immunodeficiency frequently exhibit reactivation of CMV from latency or fail to limit a primary CMV infection and often develop visceral disease. Allogeneic BMT recipients who undergo complete ablation of their immune system prior to receiving a transplant of donor bone marrow or stem cells and then receive immunosuppressive drugs posttransplant to prevent graft versus host disease (GVHD), are at especially high risk for CMV interstitial pneumonia (CMV-IP) and enteritis. Before the advent of effective antiviral drug therapy, CMV-IP and enteritis occurred in 25% and 15% of CMV seropositive BMT recipients, respectively, and more than 50% of patients developing CMV-IP had a fatal outcome (Meyers et al. 1986). The use of the antiviral drug ganciclovir either as prophylaxis to prevent CMV reactivation or as preemptive therapy at the first sign of reactivation has reduced the incidence of CMV disease in the first 100 days after BMT but a significant number of patients still develop CMV disease following discontinuation of ganciclovir due to incomplete immunologic recovery (Goodrich et al. 1993; Boeckh et al. 1996). Thus, several studies have investigated the nature of the immunologic defects that permit progressive infection with a view toward developing interventions that would correct these defects.

4.2.1.1 Nature of the Immunologic Defects

The early period after allogeneic BMT is characterized by severe deficiencies of NK cells, antibody producing B cells, and $CD4^+$ and $CD8^+$ $\alpha\beta$ T cell receptor-positive T lymphocytes. Individuals with congenital deficiencies of NK cells have an increased risk of severe and recurrent herpes virus infections and deficiencies of this subset may play a role in permitting the progression of CMV infection (Biron et al. 1989). However, NK activity appears to recover relatively early after BMT and before the peak incidence of CMV disease suggesting that other immune defects are critical for the development of CMV disease in these patients. The possibility that deficiencies of CMV-specific antibodies were responsible for progressive CMV infection was addressed in several studies in which CMV immune globulin was administered as prophylaxis for infection. This approach failed to protect BMT patients from the development of CMV disease although administration of immunoglobulin combined with ganciclovir appears to reduce the mortality rate from established CMV-IP (Emanuel et al. 1988).

The hypothesis that progressive CMV infection may be related to a quantitative deficiency of virus-specific $\alpha\beta^+$ T cell responses was supported by studies in a murine model of CMV infection (Reddehase et al. 1985). Several clinical studies of BMT recipients also identified a striking correlation between the presence of major histocompatibility complex (MHC)-restricted $\alpha\beta^+$ T cell responses to CMV and protection from the subsequent occurrence of CMV disease suggesting that recovery of the $CD8^+$ CTL response was crucial for protection of the host (Fig. 1; Quinnan et al. 1982; Reusser et al. 1991; Li et al. 1994). If the marrow donor was CMV seropositive, adoptive transfer of additional donor T cells to the recipient could potentially correct the defect in T cell immunity. However, this would require enriching the donor T cells for those reactive with CMV by in vitro culture or T cell cloning to minimize the potential for causing GVHD.

4.2.1.2 Specificity of Protective $CD8^+$ and $CD4^+$ CMV-Specific T Cells

CMV has a large genome and encodes a diverse array of antigens that could be recognized by T cells. Thus, to ensure that T cells selected for use in therapy efficiently recognized CMV-infected target cells and were representative of the immunodominant response maintained in immunocompetent hosts, it was essential to characterize the antigens recognized by CMV-specific T cells. CMV expresses its genes in discrete phases termed the immediate early (IE), early (E), and late (L) phases. To determine if CTLs preferentially recognized proteins produced at IE, E, or L stages of the replicative cycle, the timed addition of inhibitors to block viral protein or RNA synthesis was employed. Despite the application of an RNA synthesis inhibitor prior to viral infection of target cells to completely block the production of newly synthe-

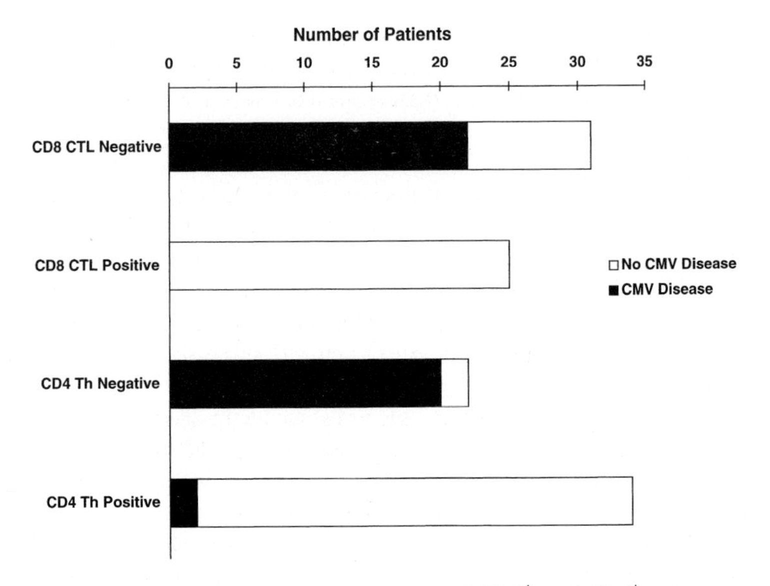

Fig. 1. Correlation of endogenous recovery of $CD8^+$ and $CD4^+$ cytomegalovirus (CMV)-specific T cell responses and protection from CMV disease after allogeneic bone marrow transplantation. Blood samples were obtained from 56 allogeneic bone marrow transplant (BMT) recipients on days 30, 60 and 90 after transplant and assayed for $CD8^+$ CMV-specific cytolytic T cell responses and $CD4^+$ CMV-specific lymphoproliferative responses as described (Reusser et al. 1991; Li et al. 1994). Patients were followed for the development of CMV enteritis and interstitial pneumonia. Patients who recovered $CD8^+$ CMV-specific T cell responses ($CD8^+$ CTL positive) or $CD4^+$ Th responses ($CD4^+$ Th positive) were protected from the development of CMV disease whereas those patients who lacked detectable $CD8^+$ CTLs ($CD8^+$ CTL negative) or $CD4^+$ Th ($CD4^+$ Th negative) had a high incidence of CMV disease

sized viral proteins, the targets were lysed by the majority of $CD8^+$ CMV-specific CTL clones (Riddell et al. 1991). This data indicated that viral proteins introduced into the cytoplasm of the infected cell after entry and uncoating were efficiently processed for recognition by CTLs. The specificity of the $CD8^+$ CMV-specific CTLs that recovered after allogeneic BMT and conferred protection from subsequent CMV disease was similarly directed at structural viral proteins (Li et al. 1994).

The matrix protein pp65 has now been identified to be the most frequent target of the immunodominant CTL response although major responses to a second matrix protein pp150 have been observed in some individuals (McLaughlin-Taylor et al. 1994; Wills et al. 1996). CMV-infected cells are targets for lysis by pp65 or pp150-specific $CD8^+$ CTLs within 1 h of viral entry and remain susceptible throughout the entire replicative cycle despite the expression of viral genes that downregulate class I MHC expression in an attempt to evade CTLs (Riddell and Greenberg 1997; Ploegh 1998). Normal CMV seropositive individuals maintain high levels of CTLs specific for pp65 and/or pp150 suggesting these effector cells are necessary to control intermittent episodes of virus reactivation.

Detailed analysis of the specificity of the $CD4^+$ CMV-specific Th response has not been performed but the data available suggest that the viral gB, pp65, and IE proteins can be targets for recognition (Forman et al. 1985; Rodgers et al. 1987; Hopkins et al. 1996). $CD4^+$ virus-specific T cells could be activated directly by virus-infected cells that express class II MHC or by neighboring class II^+ antigen-presenting cells that have endocytosed and processed viral antigens. CMV also attempts to evade recognition by $CD4^+$ T cells by expressing a gene (US2) which induces the degradation of the class II DRα and DMα chains in CMV-infected cells (Tomazin et al. 1999).

4.2.1.3 Adoptive Transfer of $CD8^+$ CMV-Specific T Cell Clones to BMT Recipients

The objectives of the initial study to evaluate adoptive immunotherapy with CMV-specific T cells were to define the safety of administering $CD8^+$ CMV-specific CTLs as prophylaxis for CMV disease and to determine the ability of the infused T cells to persist and function in vivo. The T cells were not given to patients with established CMV disease because of the potential for immunopathology if the virus has extensively infected visceral tissues.

$CD8^+$ CMV-specific CTL clones were isolated from the marrow donor, selected for specificity against structural viral proteins, and expanded to large numbers. CTLs were administered to the BMT recipient in escalating doses of $3.3\times10^7/m^2$, $1\times10^8/m^2$, $3.3\times10^8/m^2$, and $1\times10^9/m^2$ of body surface area weekly for 4 weeks beginning 28–42 days after BMT. Minor toxicities including low grade fever and

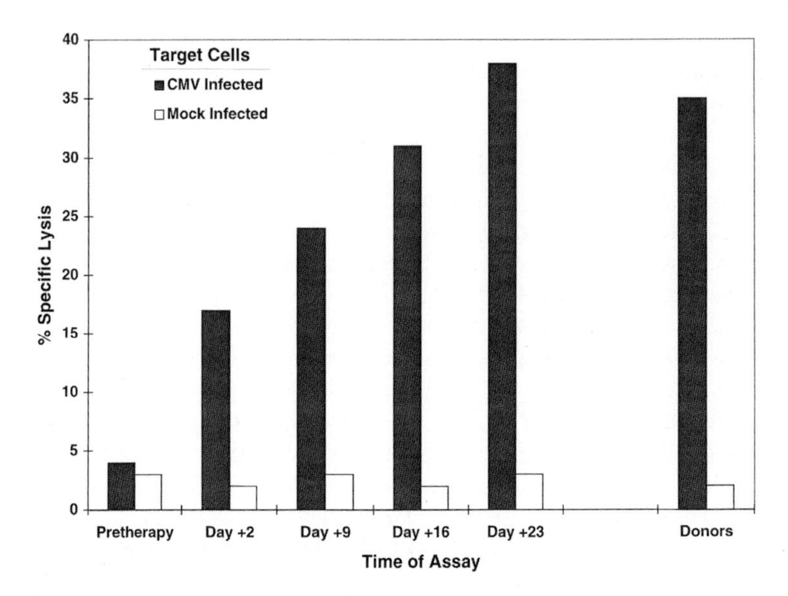

Fig. 2. Adoptive transfer of $CD8^+$ CMV-specific CTL clones restores CMV-specific immunity in allogeneic BMT recipients. $CD8^+$ CMV-specific CTL clones were administered in escalating doses each week for 4 weeks to 11 patients who lacked detectable CMV-specific CTL responses. Blood samples were obtained pretherapy and 2 days after each infusion, stimulated with autologous CMV-infected cells and assayed at an effector to target ratio of 10:1 against autologous CMV-infected and mock infected target cells. The lytic activity shown at each time point is the mean for the 11 patients. For reference, the mean lytic activity of identical cultures established from the immunocompetent donors is shown

chills were observed in 2 of the 14 patients and CTL infusions were routinely administered in the Outpatient Department.

Analysis of CMV-specific CTL reactivity in peripheral blood at multiple time points before, during, and after therapy showed that the adoptive transfer of CTLs was effective for restoring host immunity (Riddell et al. 1992; Walter et al. 1995). Eleven of the 14 patients had absent CTL responses immediately prior to the first infusion but after the 4-week treatment period exhibited CTL responses equivalent to those in the donor (Fig. 2). The contribution of infused CTLs to the responses observed in the recipients was examined in 3 patients using the unique sequences of the rearranged T cell receptor Vβ gene expressed in infused CTL clones as a marker. This analysis demonstrated that the infused CTLs persisted for at least 12 weeks after administration (Walter et al. 1995).

4.2.1.4 Requirement for CD4+ Th Cells

Virus-specific CD4+ Th responses may exert antiviral effects by direct recognition of infected cells expressing class II MHC antigens but also provide signals critical for the generation and maintenance of CD8+ CTLs. In mice deficient in CD4+ Th function, virus-specific CTLs have been shown to undergo deletion or become dysfunctional (Zajac et al. 1998). In the recipients of CD8+ CMV-specific CTLs, cytolytic responses equivalent to those in the immunocompetent marrow donor were achieved in all patients immediately after the fourth infusion but declined in the subset who failed to recover endogenous CD4+ CMV-specific Th responses (Walter et al. 1995). While this finding is consistent with a requirement for CD4+ CMV-specific Th for persistence of CTLs, this subset of patients also had developed GVHD as a consequence of the BMT and required treatment with both cyclosporine and prednisone. Thus, it was not possible to determine if the deficiency of CD4+ Th or the intense immunosuppressive drug therapy were responsible for the poor in vivo survival of transferred CTLs.

4.2.1.5 Virologic Monitoring of BMT Patients Receiving Adoptive Immunotherapy with CD8+ CMV-Specific CTLs

To provide insight into the potential efficacy of therapy, all patients enrolled in this study of adoptive immunotherapy as prophylaxis for CMV infection were followed for virus reactivation by cultures of the blood, urine, and throat. A positive culture for CMV from the throat was obtained in 1 patient before therapy and became negative after the first T cell infusion. Two patients had a positive urine culture during T cell therapy. No patient had evidence of CMV viremia and none of the 14 patients developed CMV disease (Walter et al. 1995). Thus, the phase I study has provided sufficiently encouraging evidence of antiviral activity to proceed with a larger study.

4.2.1.6 Future Studies

The administration of both $CD8^+$ CTL and $CD4^+$ Th clones to immunodeficient BMT recipients could restore both arms of the host T cell immune response and improve the persistence of the transferred $CD8^+$ CTL subset. A phase II study has been initiated to evaluate adoptive immunotherapy with both CMV-specific $CD8^+$ CTL and $CD4^+$ Th clones as a strategy for preventing CMV disease after allogeneic BMT. It is anticipated this study will also provide insight into the effects of immunosuppression on the persistence and function of transferred T cells.

4.2.2 Adoptive Immunotherapy of HIV Infection

4.2.2.1 Role of Virus-Specific T cells in the Control of HIV Replication

HIV differs from other persistent human viruses in that infection is characterized by persistent high levels of virus replication. However, experimental evidence suggests that HIV-specific T cells participate in a major way in limiting HIV replication in the host. After primary infection, the appearance of $CD8^+$ HIV-specific CTLs coincides with a reduction in viremia and return to near normal CD4 counts and in chronically infected patients, the magnitude of the HIV-specific $CD8^+$ CTL response correlates inversely with plasma viral load (Musey et al. 1997; Ogg et al. 1998). The subset of infected individuals, termed long-term non-progressors, who remain asymptomatic and retain normal CD4 counts for several years characteristically maintain strong and persistent $CD8^+$ Gag-specific CTL responses (Borrow et al. 1994; Koup 1994; Klein et al. 1995; Musey et al. 1997). The presence of $CD4^+$ HIV-specific T cells that proliferate in vitro in response to HIV antigens is also characteristic of long-term non-progressors and these responses are absent in individuals with progressive infection (Rosenberg et al. 1997).

4.2.2.2 Adoptive Transfer of $CD8^+$ HIV-Specific CTL Clones Modified with the HSV-TK Gene

Based on the findings that $CD8^+$ HIV-specific CTLs mediate antiviral activity in patients with progressive infection, we examined whether the adoptive transfer of large numbers of autologous in vitro expanded $CD8^+$ HIV-specific CTL clones could augment the partially effective host response and provide additional antiviral activity. $CD8^+$ CTLs produce inflammatory cytokines following recognition of HIV-infected cells, and augmenting the host CTL response in patients with a high virus load acutely by adoptive transfer may carry a risk of toxicity. Thus, in the initial patients the autologous $CD8^+$ HIV-specific CTL clones were modified by retrovirus-mediated gene insertion to express the herpes simplex virus thymidine kinase (HSV-TK) gene. This genetic modification would permit the transferred CTLs to be ablated by administering ganciclovir.

Six patients with progressive HIV infection were treated with infusions of TK-modified $CD8^+$ Gag-specific CTL clones in escalating doses of $1\times10^8/m^2$, $3.3\times10^8/m^2$, $1\times10^9/m^2$, and $3.3\times10^9/m^2$ at 2-week intervals (Riddell et al. 1996). Since HIV may escape CTL recognition by mutating the epitopes recognized by CTLs, the region of the *gag* gene encoding the epitope was sequenced in the patient's virus prior to therapy to ensure CTL clones specific for non-mutated epitopes were selected for use in therapy. The T cell infusions were not associated with serious toxicity although low grade fever, night sweats, and myalgias occurred commonly in the first 24–48 h after the infusion. None of the six patients required ganciclovir to ablate the transferred CTLs. Unfortunately, the in vivo persistence of the gene-modified CTLs was limited in five of the six patients by an immune response to the expressed transgene product and this interfered with conclusive assessment of safety and antiviral efficacy (Riddell et al. 1996).

4.2.2.3 Adoptive Transfer of Unmodified and LN-Modified $CD8^+$ HIV-Specific CTLs

To further examine the safety and potential antiviral activity of $CD8^+$ CTLs, a subsequent study was performed in which HIV^+ patients received three infusions of autologous $CD8^+$ HIV-specific CTL clones in doses of $3\times10^8/m^2$, $1\times10^9/m^2$, and $3.3\times10^9/m^2$ at 2-week intervals, respectively, followed by two infusions of CTLs that were genetically modified to contain the neomycin phosphotransferase gene (*neo*) to facilitate monitoring of cell persistence and localization to sites of infection (Brodie et al. 1999). The *neo*-marked CTLs were administered 1 week apart at cell doses of $1\times10^9/m^2$ and $3.3\times10^9/m^2$.

The administration of these doses of CD8$^+$ HIV-specific CTLs caused mild to moderate flu-like symptoms lasting 24–48 h and was associated with an increase in the direct killing of target cells expressing HIV Gag by peripheral blood lymphocytes (Brodie et al. 1999). The peak lytic activity was observed 1 day after the infusion of the highest cell dose, at which time *neo* -marked T cells comprised up to 3.7% of the CD8$^+$ T cells in the peripheral blood. Four days after the second infusion of *neo*-modified CTLs, a subset of patients underwent a lymph node biopsy to assess migration of CTLs to lymph node sites. Large numbers of *neo*-marked cells were detected within the lymph node architecture adjacent to cells actively replicating HIV. Thus, these results demonstrated that CD8$^+$ HIV-specific CTL responses could be safely augmented by adoptive transfer of T cell clones and that infused CTLs functioned in vivo and homed appropriately to sites of virus replication (Brodie et al. 1999).

The antiviral activity of adoptive immunotherapy was assessed by measuring plasma viral load and the frequency of circulating HIV-infected cells before and after therapy. A transient increase in plasma viral load was observed 1 day after each CTL infusion in some of the patients but returned to baseline or below within 3 days. However, a sustained reduction in plasma virus was not observed. Three patients had measurable levels of CD4$^+$ T cells that were actively replicating HIV in the peripheral blood. The frequency of these HIV-infected CD4$^+$ T cells declined dramatically over 3–4 days after each CTL infusion but returned to baseline levels by 7–10 days (Brodie et al. 1999). Accumulation of variant viruses with mutations in sequences encoding the T cell epitope was not detected suggesting the transient antiviral effect was not due to the emergence of HIV escape mutants. The return of HIV-infected cells correlated with a decline in the level of transferred CD8$^+$ CTLs in the blood suggesting that the loss of the infused CTLs permitted resurgence of virus-infected cells.

The precise mechanisms involved in the rapid loss of transferred CD8$^+$ HIV-specific CTLs have not yet been elucidated. However, insights derived from in vitro and animal model studies suggest a deficiency of IL-2 producing HIV-specific CD4$^+$ Th cells may be a critical factor. Studies of LCMV infection of CD4–/– mice have revealed an essential role for CD4$^+$ T cells in maintaining the persistence and function of CD8$^+$ virus-specific CTLs (Zajac et al. 1998). In the hu-PBL-

SCID mouse model of HIV infection, a setting in which $CD4^+$ HIV-specific Th cells are deficient, $CD8^+$ HIV-specific CTLs exerted an antiviral effect in mice challenged with HIV, but failed to persist in vivo (McKinney et al. 1999). Thus, although $CD8^+$ CTLs appear to be important effector cells for controlling HIV, their use as the sole mode of immunotherapy may be limited in most patients by the deficiency of $CD4^+$ HIV-specific Th.

4.2.2.4 Strategies to Provide Th Activity in HIV-Infected Patients

Several strategies are being developed to correct deficient Th function in HIV-infected hosts and improve the persistence and function of adoptively transferred CTLs. The administration of IL-2 with transferred CTLs improves CTL persistence and antiviral activity in animal models (Reddehase et al. 1987), and is now being evaluated in HIV-infected patients. A more physiologic approach to deliver IL-2 to activated CTLs may be to genetically engineer $CD8^+$ CTLs to have a Th-independent phenotype. One strategy which has evolved from an improved understanding of IL-2 signal transduction utilizes the expression of chimeric cytokine receptors consisting of the extracellular domains of the GM-CSF α and β receptors fused to the intracellular domains of the IL-2 γ and β chains, respectively (Evans et al. 1999). $CD8^+$ CTLs produce GM-CSF after antigen stimulation and engagement of the chimeric GM-CSF/IL-2 receptor chains by GM-CSF induces heterodimerization of the intracellular IL-2 β and γ chains and activates the downstream IL-2 signaling machinery (Nelson et al. 1994). An alternative strategy would be to restore or augment HIV-specific $CD4^+$ T cell responses by adoptive transfer. Recent studies have demonstrated that $CD4^+$ Th responses to HIV antigens can be detected in HIV-infected individuals (Pitcher et al. 1999). Thus, it may be feasible to isolate these cells in vitro, introduce “intracellular immunization” genes to render the cells resistant to HIV, and then expand and reinfuse them into the patient to provide an amplified and sustained HIV-specific Th response.

4.3 Immunotherapy of Malignancy: T Cell Therapy of Leukemia after Allogeneic BMT

The recent identification of antigens expressed by human tumor cells has given new impetus to efforts to treat malignant disease with immunotherapy. Much of the effort has focused on the treatment of melanoma with autologous T cells, however, success has also been achieved in treating relapsed leukemia after allogeneic BMT (van Rhee and Kolb 1995; Rosenberg 1999). Transplantation of unmodified bone marrow or stem cells from an allogeneic donor who is identical with the recipient at the MHC provides a graft versus leukemia (GVL) effect which is not seen with transplantation of T cell-depleted allogeneic stem cell preparations or with transplantation using a syngeneic donor (Goldman et al. 1988; Horowitz et al. 1990). The GVL effect associated with allogeneic transplant is linked with GVHD and is presumed to be due to differences in minor histocompatibility antigens which are encoded by polymorphic genetic loci that differ between donor and recipient and are presented to T cells as peptide fragments associated with class I and class II MHC molecules (Goulmy 1997a).

4.3.1 Tissue-Restricted Minor H Antigens as Targets for GVL Therapy

GVHD and GVL responses are usually linked but a GVL effect can be observed even in the absence of clinical GVHD (Horowitz et al. 1990). This may be explained by the differential expression of minor H antigens on host tissues (de Bueger et al. 1992). T cells specific for some minor H antigens recognize hematopoietic cells including leukemic progenitors but not cells derived from other tissues such as skin. Such tissue restricted minor H antigens could potentially be targets for adoptive immunotherapy or vaccination to promote a GVL response without causing GVHD (Arnold et al. 1995; Warren et al. 1998a).

T cells reactive with human minor H antigens and potentially involved in GVHD and GVL responses have been isolated from patients receiving MHC-matched BMT. The antigens recognized by these T cells can be broadly grouped into those encoded by the Y chromosome and those encoded by autosomal genes. T cells reactive with

Y chromosome-encoded (H-Y) antigens are only isolated in settings in which the patient and donor are sex mismatched. H-Y-specific T cells restricted by HLA A1, A2, B7, and B8 have been isolated and characterized (Goulmy 1997a; Warren et al. 1998b). The human SMCY gene was shown to encode the H-Y peptides presented by HLA A2 and HLA B7 (Wang et al. 1995; Meadows et al. 1997). SMCY is expressed in all tissues, and cells of both hematopoietic and non-hematopoietic origin are lysed by SMCY-specific T cells (Goulmy 1997a). Thus, SMCY is not an optimal target for T cell therapy to induce a selective GVL response. Recently, $CD8^+$ T cell clones specific for a novel H-Y antigen, which is presented associated with HLA B8, have been isolated by our group after transplantation of marrow from a female donor into a male recipient. These H-Y-specific CTLs lysed hematopoietic cells including leukemic blasts but not skin fibroblasts obtained from male HLA $B8^+$ donors (Warren et al. 1998a,b). Using HLA $B8^+$ cell lines (provided by David Page and Elizabeth Simpson) derived from males with deletions of portions of the Y chromosome as target cells, the gene encoding the HLA B8-restricted H-Y antigen was shown to be distinct from SMCY. Further analysis of this novel Y chromosome gene should delineate its potential role as a target for GVL therapy.

$CD8^+$ CTLs specific for a large number of minor H antigens encoded by autosomal genes have been isolated and characterized by several groups. Goulmy (1997b) and van Els et al. (1992) have characterized $CD8^+$ CTL clones which define seven minor H antigens designated HA-1 to HA-7. CTLs specific for the HA-3, -4, -5, -6, and -7 antigens lysed hematopoietic cells, endothelial cells, epithelial cells, and fibroblasts suggesting the epitopes are derived from ubiquitously expressed genes and are not optimal targets for GVL therapy (de Bueger et al. 1992; Goulmy 1997a). However, T cells specific for HA-1 and HA-2 fail to lyse non-hematopoietic cells and have been proposed as potential targets for T cell therapy (Mutis et al. 1999).

In a recent study by our group, $CD8^+$ CTL clones specific for 19 minor H antigens encoded by autosomal genes were isolated (Warren et al. 1998b). Ten of these minor H antigens were presented by HLA A2 or B7 but were distinguished from HA-1, -2, -4, -5, -6, and -7 by their tissue expression and/or the frequency of the allele encoding the minor H antigen in the population. The other nine minor H antigens were presented by either HLA A3, A11, A29, B44, B53, or Cw7 (Warren et al. 1998b). Fourteen of these 19 minor H antigens were expressed in hematopoietic cells including AML, ALL, and CLL cells but not in skin fibroblasts and are being pursued as targets for GVL therapy (Warren et al. 1998b).

Dolstra et al. (1997, 1999) have isolated a $CD8^+$ CTL clone specific for a minor H antigen denoted HB-1 which is presented by HLA B44. This antigen is of considerable interest as a GVL target because its expression is limited to transformed B lymphocytes. Thus, HB-1-specific CTLs lyse EBV-transformed B cells and B-ALL cells but not skin fibroblasts, T cells, and monocytes.

4.3.2 Expression of Minor H Antigens on Leukemic Progenitor Cells

Several approaches have been used to assess recognition of leukemic cells by minor H antigen-specific CTLs including assessment of the ability of CTLs to lyse leukemic blasts in vitro or to inhibit leukemic colony formation in soft agar. These studies demonstrate that minor H antigens are expressed on at least some of the leukemic blast population but do not ascertain whether the leukemic stem cell is a target for CTLs (Falkenberg et al. 1991; van der Harst et al. 1994; Warren et al. 1998b). The inoculation of human AML into immunodeficient mice has defined a putative leukemic stem cell which is required for engraftment (Bonnet and Dick 1997). This SCID leukemia initiating cell (SLIC) is $CD34^+$ $CD38^-$ but present in very low frequency ($/10^5$ cells) in the blast population (Lapidot et al. 1994; Bonnet and Dick 1997). Recent studies in our laboratory show that the engraftment of AML can be specifically prevented by minor H antigen-specific CTL clones demonstrating that the rare SLIC expresses minor H antigens and can be eliminated by CTLs (Bonnet et al. 1999). Engraftment of stem cells derived from a donor that does not encode the minor H allele was not affected by CTLs suggesting that administering minor H antigen-specific T cells to BMT recipients should not interfere in a non-specific fashion with engraftment of donor hematopoietic cells.

4.3.3 Identification of genes encoding minor H antigens

Identification of the genes encoding minor H antigens will facilitate analysis of the distribution of antigen expression in vivo and selection of antigens to target for GVL therapy. Three strategies are being used by various laboratories to identify genes encoding minor H antigens. The first strategy utilizes cloning techniques to define the chromosome location of the gene using a panel of cells obtained from large families that have been extensively mapped for hundreds of polymorphic loci. The gene encoding an HLA B7 restricted minor H antigen has been localized with this technique to chromosome 22 near the platelet-derived growth factor and IL-2Rβ loci (Gubarev et al. 1996). At the present time, considerable effort is still required to identify the specific gene and to derive the sequence of the antigenic epitope, however, the rapid progress in sequencing the human genome will improve the utility of this approach.

The second approach relies on the elution of peptides from cell surface MHC molecules, separation of the fraction containing the minor H antigen peptide using biochemical techniques, and derivation of the sequence of the peptide using mass spectrometry (Hunt et al. 1992). The nucleotide sequence encoding the epitope can be deduced from the amino acid sequence of the peptide and databases containing human DNA sequences searched to determine if the epitope is encoded by a known gene. The peptide elution method has been used to define the HA-1 and HA-2 peptides which have been suggested as potential targets of a selective GVL response. The amino acid sequence of HA-2 closely matches but is not identical to that of a class I myosin gene (den Haan et al. 1995). The failure to precisely assign the amino acid sequence of HA-2 to a known gene has precluded a molecular analysis of the expression of this antigen in primary tissues. The deduced DNA sequence encoding HA-1 was found in a gene derived from an AML cell line and present in at least two alleles that differ at only a single amino acid in the nonamer epitope recognized by HA-1-specific CTLs (den Haan et al. 1998).

The third approach to identify genes encoding $CD8^+$ CTL defined minor H antigens uses a cDNA expression cloning strategy pioneered by Boon et al. (1994). This involves cotransfection of antigen-negative target cells with pools of cDNA from a library derived from an antigen-positive cell and a cDNA encoding the class I MHC restricting allele. The proteins expressed by the transfected cDNAs will be processed and presented with the introduced class I MHC molecule. Recognition of the transfected cells by T cells can be assessed by cytokine release and cDNA pools inducing a positive response are then subdivided and screened until a single cDNA encoding the antigen is identified. This strategy is complementary to the peptide elution approach and is being used by our group to identify genes encoding minor H antigens considered to be potential candidates for GVL therapy.

4.3.4 Strategies for Clinical Application

Relapse remains a major obstacle to a successful outcome after allogeneic BMT for acute leukemia (Appelbaum 1997). Thus, we are now conducting a study in which T cell clones specific for recipient minor H antigens are prospectively isolated for later use in immunotherapy. Clones with reactivity for recipient hematopoietic cells including leukemic blasts but not for non-hematopoietic cells are being stored and, if the patient relapses, these cells would be expanded for infusion. It is anticipated a complete molecular analysis of the target antigen will not always be available and there will be some risk of GVHD. Thus, in this initial study the HSV-TK gene will be introduced into the T cell clones to permit their ablation by the administration of ganciclovir. This strategy has been used successfully to ameliorate GVHD occurring following adoptive immunotherapy with unselected polyclonal donor lymphocytes modified with TK (Bonini et al. 1997).

4.4 Conclusions

Several human diseases have been identified in which disease progression is associated with deficiencies of antigen-specific T cell responses. Improvements in T cell culture methodology have made it feasible to propagate T cells to numbers sufficient for immunotherapy of humans and the availability of molecular approaches to characterize antigens encoded by pathogens or tumor cells has facilitated isolation and identification of appropriate effector T cells. Early clinical results have been encouraging and it is anticipated that future studies will provide new insights into the potential efficacy of adoptive immunotherapy with T cells in human viral and malignant diseases.

References

Appelbaum FR (1997) Allogeneic hematopoietic stem cell transplantation for acute leukemia. Semin Oncol 24:114–123

Arnold D, Faath S, Rammensee H, Schild H (1995) Cross-priming of minor histocompatibility antigen-specific cytotoxic T cells upon immunization with the heat shock protein gp96. J Exp Med 182:885–889

Biron CA, Byron KS, Sullivan JL (1989) Severe herpesvirus infections in an adolescent without natural killer cells. N Engl J Med 320:1731–1735

Boeckh MRS, Cunningham T, Myerson D, Flowers M, Bowden R (1996) Increased risk of late CMV infection and disease in allogeneic marrow transplant recipients after ganciclovir prophylaxis is due to a lack of CMV-specific T cell responses. Blood (suppl):1195A

Bonini C, Ferrari G, Verzeletti S, Servida P, Zappone E, Ruggieri L, et al (1997) HSV-TK gene transfer into donor lymphocytes for control of allogeneic graft-versus-leukemia. Science 276:1719–1724

Bonnet D, Dick JE (1997) Human acute myeloid leukemia is organized as a hierarchy that originates from a primitive hematopoietic cell. Nat Med 3:730–737

Bonnet D, Warren EH, Greenberg PD, Dick J, Riddell SR (1999) CD8+ minor histocompatibility antigen-specific cytotoxic T lymphocyte clones eliminate AML skin cells. Proc Natl Acad Sci USA 3:730–737

Boon T, Cerottini JC, Van den Eynde B, Bruggen P van der, Van Pel A (1994) Tumor antigens recognized by T lymphocytes. Annu Rev Immunol 12:337–365

Borrow P, Lewicki H, Hahn BH, Shaw GM, Oldstone MB (1994) Virus-specific CD8+ cytotoxic T-lymphocyte activity associated with control of viremia in primary human immunodeficiency virus type 1 infection. J Virol 68:6103–6110

Brodie SJ, Lewinsohn DA, Patterson BK, Jiyamapa D, Krieger J, Corey L, et al (1999) In vivo migration and function of transferred HIV-1-specific cytotoxic T cells. Nat Med 5:34–41

Bueger M de, Bakker A, Van Rood JJ, Van der Woude F, Goulmy E (1992) Tissue distribution of human minor histocompatibility antigens. Ubiquitous versus restricted tissue distribution indicates heterogeneity among human cytotoxic T lymphocyte-defined non-MHC antigens. J Immunol 149:1788–1794

Dolstra H, Fredrix H, Preijers F, Goulmy E, Figdor CG, Witte TM de, et al (1997) Recognition of a B cell leukemia-associated minor histocompatibility antigen by CTL. J Immunol 158:560–565

Dolstra H, Fredrix H, Maas F, Coulie PG, Brasseur F, Mensink E, et al (1999) A human minor histocompatibility antigen specific for B cell acute lymphoblastic leukemia. J Exp Med 189:301–308

Els CA van, D'Amaro J, Pool J, Blokland E, Bakker A, Elsen PJ van, et al (1992) Immunogenetics of human minor histocompatibility antigens: their polymorphism and immunodominance. Immunogenetics 35:161–165

Emanuel D, Cunningham I, Jules-Elysee K, Brochstein JA, Kernan NA, Laver J, et al (1988) Cytomegalovirus pneumonia after bone marrow transplantation successfully treated with the combination of ganciclovir and high-dose intravenous immune globulin. Ann Intern Med 109:777–782

Evans LS, Witte PR, Feldhaus AL, et al (1999) Expression of a GM-CSF/IL-2 chimeric receptor in human CTL clones results in GM-CSF dependent growth. Hum Gene Ther 10:1942–1951

Falkenburg JH, Goselink HM, Harst D van der, Luxemburg-Heijs SA van, Kooy-Winkelaar YM, Faber LM, et al (1991) Growth inhibition of clonogenic leukemic precursor cells by minor histocompatibility antigen-specific cytotoxic T lymphocytes. J Exp Med 174:27–33

Forman SJ, Zaia JA, Clark BR, Wright CL, Mills BJ, Pottathil R, et al (1985) A 64,000 dalton matrix protein of human cytomegalovirus induces in vitro immune responses similar to those of whole viral antigen. J Immunol 134:3391–3395

Goldman JM, Gale RP, Horowitz MM, Biggs JC, Champlin RE, Gluckman E, et al (1988) Bone marrow transplantation for chronic myelogenous leukemia in chronic phase. Increased risk for relapse associated with T-cell depletion. Ann Intern Med 108:806–814

Goodrich JM, Bowden RA, Fisher L, Keller C, Schoch G, Meyers JD (1993) Ganciclovir prophylaxis to prevent cytomegalovirus disease after allogeneic marrow transplant. Ann Intern Med 118:173–178

Goulmy E (1997a) Human minor histocompatibility antigens: new concepts for marrow transplantation and adoptive immunotherapy. Immunol Rev 157:125–140

Goulmy E (1997b) Minor histocompatibility antigens: from T cell recognition to peptide identification (editorial). Hum Immunol 54:8–14

Gubarev MI, Jenkin JC, Leppert MF, Buchanan GS, Otterud BE, Guilbert DA, et al (1996) Localization to chromosome 22 of a gene encoding a human minor histocompatibility antigen. J Immunol 157:5448–5454

Haan JM den, Sherman NE, Blokland E, Huczko E, Koning F, Drijfhout JW, et al (1995) Identification of a graft versus host disease-associated human minor histocompatibility antigen. Science 268:1476–1480

Haan JM den, Meadows LM, Wang W, Pool J, Blokland E, Bishop TL, et al (1998) The minor histocompatibility antigen HA-1: a diallelic gene with a single amino acid polymorphism. Science 279:1054–1057

Harst D van der, Goulmy E, Falkenburg JH, Kooij-Winkelaar YM, Luxemburg-Heijs SA van, Goselink HM, et al (1994) Recognition of minor histocompatibility antigens on lymphocytic and myeloid leukemic cells by cytotoxic T-cell clones. Blood 83:1060–1066

Hopkins JI, Fiander AN, Evans AS, Delchambre M, Gheysen D, Borysiewicz LK (1996) Cytotoxic T cell immunity to human cytomegalovirus glycoprotein B. J Med Virol 49:124–131

Horowitz MM, Gale RP, Sondel PM, Goldman JM, Kersey J, Kolb HJ, et al (1990) Graft-versus-leukemia reactions after bone marrow transplantation. Blood 75:555–562

Hunt DF, Henderson RA, Shabanowitz J, Sakaguchi K, Michel H, Sevilir N, et al (1992) Characterization of peptides bound to the class I MHC molecule HLA-A2.1 by mass spectrometry. Science 255:1261–1263

Klein MR, Baalen CA van, Holwerda AM, Kerkhof Garde SR, Bende RJ, Keet IP, et al (1995) Kinetics of Gag-specific cytotoxic T lymphocyte responses during the clinical course of HIV-1 infection: a longitudinal analysis of rapid progressors and long-term asymptomatics. J Exp Med 181:1365–1372

Koup RA (1994) Virus escape from CTL recognition. J Exp Med 180:779–782

Lapidot T, Sirard C, Vormoor J, Murdoch B, Hoang T, Caceres-Cortes J, et al (1994) A cell initiating human acute myeloid leukaemia after transplantation into SCID mice. Nature 367:645–648

Li CR, Greenberg PD, Gilbert MJ, Goodrich JM, Riddell SR (1994) Recovery of HLA-restricted cytomegalovirus (CMV)-specific T-cell responses after allogeneic bone marrow transplant: correlation with CMV disease and effect of ganciclovir prophylaxis. Blood 83:1971–1979

McKinney DM, Lewinsohn DA, Riddell SR, et al (1999) The antiviral activity of HIV-specific CD8+ CTL clones is limited by elimination due to encounter with HIV-infected targets. J Immunol 163:861–867

McLaughlin-Taylor E, Pande H, Forman SJ, Tanamachi B, Li CR, Zaia JA, et al (1994) Identification of the major late human cytomegalovirus matrix protein pp65 as a target antigen for CD8+ virus-specific cytotoxic T lymphocytes. J Med Virol 43:103–110

Meadows L, Wang W, Haan JM den, Blokland E, Reinhardus C, Drijfhout JW, et al (1997) The HLA-A*0201-restricted H-Y antigen contains a posttranslationally modified cysteine that significantly affects T cell recognition. Immunity 6:273–281

Meyers JD, Flournoy N, Thomas ED (1986) Risk factors for cytomegalovirus infection after human marrow transplantation. J Infect Dis 153:478–488

Musey L, Hughes J, Schacker T, Shea T, Corey L, McElrath MJ (1997) Cytotoxic-T-cell responses, viral load, and disease progression in early human immunodeficiency virus type 1 infection. N Engl J Med 337:1267–1274

Mutis T, Verdijk R, Schrama E, Esendam B, Brand A, Goulmy E (1999) Feasibility of immunotherapy of relapsed leukemia with ex vivo-generated cytotoxic T lymphocytes specific for hematopoietic system-restricted minor histocompatibility antigens. Blood 93:2336–2341

Nelson BH, Lord JD, Greenberg PD (1994) Cytoplasmic domains of the interleukin-2 receptor beta and gamma chains mediate the signal for T-cell proliferation. Nature 369:333–336

Ogg GS, Jin X, Bonhoeffer S, Dunbar PR, Nowak MA, Monard S, et al (1998) Quantitation of HIV-1-specific cytotoxic T lymphocytes and plasma load of viral RNA. Science 279:2103–2106

Pitcher CJ, Quittner C, Peterson DM, Connors M, Koup RA, Maino VC, et al (1999) HIV-1-specific CD4+ T cells are detectable in most individuals with active HIV-1 infection, but decline with prolonged viral suppression. Nat Med 5:518–525

Ploegh HL (1998) Viral strategies of immune evasion. Science 280:248–253

Quinnan GV Jr, Kirmani N, Rook AH, Manischewitz JF, Jackson L, Moreschi G, et al (1982) Cytotoxic T cells in cytomegalovirus infection: HLA-restricted T-lymphocyte and non-T-lymphocyte cytotoxic responses correlate with recovery from cytomegalovirus infection in bone-marrow-transplant recipients. N Engl J Med 307:7–13

Reddehase MJ, Weiland F, Munch K, Jonjic S, Luske A, Koszinowski UH (1985) Interstitial murine cytomegalovirus pneumonia after irradiation: characterization of cells that limit viral replication during established infection of the lungs. J Virol 55:264–273

Reddehase MJ, Mutter W, Munch K, Buhring HJ, Koszinowski UH (1987) CD8-positive T lymphocytes specific for murine cytomegalovirus immediate-early antigens mediate protective immunity. J Virol 61:3102–3108

Reusser P, Riddell SR, Meyers JD, Greenberg PD (1991) Cytotoxic T-lymphocyte response to cytomegalovirus after human allogeneic bone marrow transplantation: pattern of recovery and correlation with cytomegalovirus infection and disease. Blood 78:1373–1380

Rhee F van, Kolb HJ (1995) Donor leukocyte transfusions for leukemic relapse. Curr Opin Hematol 2:423–430

Riddell SR, Greenberg PD (1995) Principles for adoptive T cell therapy of human viral diseases. Annu Rev Immunol 13:545–586

Riddell SR, Greenberg PD (1997) T cell therapy of human CMV and EBV infections in immunocompromised hosts. Rev Med Virol 7:181–192

Riddell SR, Rabin M, Geballe AP et al (1991) Class I MHC-restricted cytotoxic T lymphocyte recognition of cells infected with human cytomega-

lovirus does not require endogenous viral gene expression. J Immunol 146:2795–2804
Riddell SR, Watanabe KS, Goodrich JM, Li CR, Agha ME, Greenberg PD (1992) Restoration of viral immunity in immunodeficient humans by the adoptive transfer of T cell clones. Science 257:238–241
Riddell SR, Elliott M, Lewinsohn DA, Gilbert MJ, Wilson L, Manley SA, et al (1996) T-cell mediated rejection of gene-modified HIV-specific cytotoxic T lymphocytes in HIV-infected patients. Nat Med 2:216–223
Rodgers B, Borysiewicz L, Mundin J, Graham S, Sissons P (1987) Immunoaffinity purification of a 72 K early antigen of human cytomegalovirus: analysis of humoral and cell-mediated immunity to the purified polypeptide. J Gen Virol 68:2371–2378
Rosenberg SA (1999) A new era for cancer immunotherapy based on the genes that encode cancer antigens. Immunity 10:281–287
Rosenberg ES, Billingsley JM, Caliendo AM, Boswell SL, Sax PE, Kalams SA, et al (1997) Vigorous HIV-1-specific CD4+ T cell responses associated with control of viremia. Science 278:1447–1450
Tomazin R, Boname J, Hegde NR, Lewinsohn DA, Altschuler Y, Jones TR, et al (1999) Cytomegalovirus US2 destroys two components of the MHC class II pathway preventing recognition by CD4+ T cells. Nat Med 5:1039–1043
Walter EA, Greenberg PD, Gilbert MJ, Finch RJ, Watanabe KS, Thomas ED, et al (1995) Reconstitution of cellular immunity against cytomegalovirus in recipients of allogeneic bone marrow by transfer of T-cell clones from the donor. N Engl J Med 333:1038–1044
Wang W, Meadows LR, Haan JM den, Sherman NE, Chen Y, Blokland E, et al (1995) Human H-Y: a male-specific histocompatibility antigen derived from the SMCY protein. Science 269:1588–1590
Warren EH, Gavin M, Greenberg PD, Riddell SR (1998a) Minor histocompatibility antigens as targets for T-cell therapy after bone marrow transplantation. Curr Opin Hematol 5:429–433
Warren EH, Greenberg PD, Riddell SR (1998b) Cytotoxic T-lymphocyte-defined human minor histocompatibility antigens with a restricted tissue distribution. Blood 91:2197–2207
Wills MR, Carmichael AJ, Mynard K, Jin X, Weekes MP, Plachter B, et al. (1996) The human cytotoxic T-lymphocyte (CTL) response to cytomegalovirus is dominated by structural protein pp65: frequency, specificity, and T-cell receptor usage of pp65-specific CTL. J Virol 70:7569–7579
Zajac AJ, Blattman JN, Murali-Krishna K, Sourdive DJ, Suresh M, Altman JD, et al (1998) Viral immune evasion due to persistence of activated T cells without effector function. J Exp Med 188:2205–2213

5 Immune Monitoring in Cancer Immunotherapy

P. Romero, M. J. Pittet, D. Valmori, D. E. Speiser, V. Cerundolo, D. Liénard, F. Lejeune, J.-C. Cerottini

5.1 Introduction

The one-century-old idea of turning the immense resources of the immune system against cancer cells as a tool in the treatment of cancer patients has been reinvigorated with the identification of the molecular targets recognized by tumor-specific cytolytic T lymphocytes (CTLs). Indeed, over 40 different antigenic peptides, derived from mostly non-mutated and normally expressed proteins, have been defined that mimic an equal number of CTL-defined tumor antigens. The possibility to chemically synthesize large quantities of well-defined antigenic peptides together with significant progress made in the understanding of in vivo induction of specific CTL responses, have paved the way to new approaches for the design of therapeutic cancer vaccines.

The goals of peptide-based immunotherapy of cancer are to assess the safety of injection of synthetic antigenic peptides in conjunction with adjuvants, to evaluate the immunogenicity of peptide-based vaccines in terms of their ability to induce specific CTL responses, and to determine their impact on tumor progression. Thus, in contrast to classic clinical trials of immunotherapy, this new approach targeted to well-defined tumor antigens is now aimed at primarily measuring a biological response. The expectation is that discrete changes in the levels of antigen-specific CTL responses translate into measurable anti-tumor effects.

These new ideas have been swiftly tested by a receptive community of tumor immunologists and oncologists. As the majority of CTL-defined tumor antigens have been defined for melanoma, it is understandable that the majority of phase I clinical trials have been performed in metastatic melanoma patients. The initial results, although encouraging, have posed new questions and new challenges. Perhaps one of the major challenges ahead concerns the means to monitor the evolution of antigen-specific CTL responses during the course of vaccination. Recent progress in this field has yielded new tools that may allow the direct quantitation of antigen-specific lymphocytes as well as their characterization in terms of activation and/or differentiation state.

5.2 First Clinical Trials of Antigen-Specific Immunotherapy

As mentioned above, small groups of mostly metastatic melanoma patients have already been immunized with synthetic antigenic peptides either alone or in combination with adjuvants (Table 1). In one trial, repeated subcutaneous injections of relatively small quantities of the MAGE-3 171–179 nonapeptide in saline were associated with an overall tumor response rate of 28%. However, no specific CTLs could be detected in peripheral blood lymphocytes (PBL) from responding patients (Marchand et al. 1999). In another trial, clear evidence of induction of specific CTL responses, as measured by interferon gamma (IFN-γ) release by PBL in vitro stimulated with the immunizing peptide, was shown in patients repeatedly injected subcutaneously with the Melan-A/MART-1 nonapeptide in incomplete Freund's adjuvant (IFA). However, no evidence of tumor response was found in this group of vaccinated patients (Cormier et al. 1997).

Table 1. Clinical trials with peptide-based cancer vaccines

Peptide	HLA-	Adjuvant	CTLs[a]	Tumor response	Reference
MAGE-3$_{168–176}$	A1	None	Not detected	7/25 **(28%)**	Marchand et al. (1999)
Melan-A$_{26–35}$	A2	None or GM-CSF	3/6		Jäger et al. (1996)
Tyrosinase$_{1–9}$	A2	None or GM-CSF	2/6		Idem
Tyrosinase$_{368–376}$	A2	None or GM-CSF			Idem
gp100$_{280–288}$	A2	None or GM-CSF	0/6		Idem
gp100$_{457–466}$	A2	None or GM-CSF	0/6	3/3	Idem
Influenza matrix$_{58–66}$	A2	None or GM-CSF			Idem
Melan-A$_{27–35}$	A2	IFA	**12/18**	0/23	Cormier et al. (1997)
gp100$_{209–217}$	A2	IFA	2/8	1/19	Rosenberg et al. (1998)
gp100$_{209–217}$ (T210 M)	A2	IFA	**10/11**	0/11	Idem
gp100$_{209–217}$ (T210 M)	A2	IL2+IFA	3/19	8/19 **(42%)**	
				5/12 **(42%)**	Idem
Various, tumor lysates	A1, A2, others	DC	N.T. (DTH, instead)	5/16 **(31%)**	Nestle et al. (1998)
MUC-1 peptide conjugated to KLH	A1,A2, A11	Detox-B	**7/11**	Not reported	Reddish et al. (1998)

CTLs, cytolytic T lymphocytes; IFA, incomplete Freund's adjuvant; GM-CSF, granulocyte-macrophage colony stimulating factor; DC, dendritic cell; N.T., not tested; DTH, delayed-type hypersensitivity
[a]Peptide-specific CTL activity measured before and after vaccination. The *numbers* indicate the fraction of individuals with an increased CTL activity after completion of the vaccination schedule over the total number of vaccinated and evaluated individuals. In *boldface type* are indicated the apparently successful CTL induction since the majority of individuals had a measurable response. The methods used to assess the response are summarized in Table 2

Several phase I trials of immunization with gp100-derived antigenic peptides have been reported. Initially, groups of patients immunized with two such antigenic peptides, which are recognized by HLA-A2-restricted CTLs, showed marginal antigen-specific T cell responses and no tumor responses (Salgaller et al. 1996). In a subsequent trial, peptide-specific responses in the majority of vaccinated patients could be detected when a modified gp100 209–217 T210 M peptide analogue was injected repeatedly in conjunction with IFA instead of the parental peptide (Rosenberg et al. 1998). Thus, these results eloquently extended to humans previous observations made in vitro and in animal models that peptide analogues with improved binding to MHC class I mole-

Table 2. Procedures used to monitor the antigen-specific CTL response in clinical trials of tumor peptide-based vaccination

Vaccine	Cells	In vitro stimulation[a]	Assay	Reference
MAGE-$3_{168-176}$ (HLA-A1-restriced)	CD8+ purified from frozen PBL	Weekly ×3	^{51}Cr-release	Marchand et al. (1999)
Melan-A/MART-1_{27-35} in IFA	Frozen PBL	Weekly ×3	^{51}Cr-release IFN-γ release	Cormier et al. (1997)
Melan-A/MART-1, tyrosinase and gp100-derived HLA-A2-restricted peptides alone i.d. followed by peptides i.d. and GM-CSF s.c.	Frozen PBL	Weekly ×2	^{51}Cr-release "DTH" to i.d. peptide injection	Jäger et al. (1996)
gp100 peptides in IFA	Fresh or frozen PBL	Once or twice	^{51}Cr-release IFN-γ release	Salgaller et al. (1996)
gp100 peptides in IFA ± systemic high dose rIL-2	Frozen PBL	Once, tested at day 11 post-stimulation	IFN-γ release	Rosenberg et al. (1998)
Autologous dendritic cells pulsed with HLA-A1 and/or HLA-A2-restricted peptides, or pulsed with tumor lysate	Biopsy infiltrating lymphocytes	Weekly ×2	^{51}Cr-release "DTH" to i.d. peptide injection	Nestle et al (1998)

IFN-γ, interferon gamma; PBL, peripheral blood lymphocytes
[a] All groups used an indirect approach based on in vitro expansion of antigen-specific CTL precursors driven by autologous antigen-presenting cells pulsed with the immunizing peptide or, sometimes, with an antigenic peptide analogue with enhanced antigenicity

cules displayed an immunogenicity considerably greater than that of their parental counterparts (Parkhurst et al. 1996; Bakker et al. 1997; Men et al. 1999).

In spite of the apparent increased immunogenicity of the gp100 peptide analogue injected in IFA, no tumor responses were recorded. An extension of the same trial that included concomitant systemic administration of high dose IL-2 appeared to have two important effects. While the proportion of peptide analogue vaccinated patients with detectable specific CTLs dropped significantly, there was a significant tumor response rate (about 42%; Rosenberg et al. 1998). Again, these results suggest a paradoxical inverse correlation between the specific CTL response that can be measured in the circulating lymphocyte compartment and tumor regression upon peptide vaccination. However, a closer look at the data on assessment of the CTL response (Table 2) reveals that a major limitation in these initial trials of peptide-based immunotherapy is the shortcomings of the methods used to measure specific CTLs. These are not only indirect but also relatively insensitive. Moreover, the source of lymphocytes was peripheral blood in all the studies. Thus, no information is available on the magnitude of the CTL response in other compartments of the immune system and, more importantly, at the tumor sites.

5.3 Traditional Methods of Monitoring CTL Responses

Since antigen recognition and effector function by CTLs involves the formation of tight cell–cell conjugates, direct assaying of $CD8^+$ T cell responses had remained impossible until recently. Binding assays were only of limited value because of the intrinsic low affinity/avidity of the clonotypic $\alpha\beta$ T cell receptor (TCR) for the cognate major histocompatibility complex (MHC) class I glycoprotein/peptide complex (the antigen ligand). The standard chromium release assay remains one of the tests most widely used but it is only semi-quantitative and relies on the measurement of a function (cytolysis dependent on the ability to degranulate and thus release lytic protein mediators) of a population of cells (for review see Romero et al. 1998a).

Efforts were also made to identify TCR α/β gene segments associated with the recognition of a defined antigen ligand thus allowing to monitor specific CTLs by molecular means such as reverse transcription-polymerase chain reaction (RT-PCR). However, this approach proved to be disappointing as many CTL responses against single antigen ligands were found to be highly diverse in terms of the repertoire of specific TCRs (Casanova et al. 1992; Yanagi et al. 1992; Cole et al. 1994; Romero et al. 1995, 1997).

The most quantitative assay system was the limiting dilution analysis (LDA). This protocol based on cloning in multiple microcultures requires the CTL precursors to undergo a minimum of 10–11 cycles of replication over a 1-week period prior to determining CTL activity by chromium release for individual culture wells (McMichael and O'Callaghan 1998). The LDA approach suffer from major limitations such as high variability and failure to measure activated CTLs (CTLe) during the peak of a CTL response against acute viral infections (Green and Scott 1994). A likely explanation was that CTLe further stimulated under limiting dilution conditions undergo apoptosis.

5.4 New Approaches to Monitoring CTL Responses

5.4.1 Enumeration of Cytokine-Releasing Lymphocytes

This is another indirect approach that, like the LDA, involves measuring an effector function of $CD8^+$ T cells, in this case the production of a cytokine. Several cytokines have been proposed but IFN-γ remains the cytokine most commonly used. Unlike the LDA, this method possesses the advantage that IFN-γ production can be detected at the single cell level and thus there is no need for antigen-specific T cells to expand in order to be detected by the assay. An additional advantage of this method of T cell counting is the shortened culture time which is of 20–48 h as compared to 1–2 weeks for LDA.

Two experimental procedures have been designed to enumerate IFN-γ-producing T cells. In the enzyme-linked immunospot (ELISPOT) assay, cell suspensions are placed in nitrocellulose-lined microtiter plates that have been previously coated with an anti-IFN-γ antibody and cultured for 1–2 days in the presence of antigen. In this way, IFN-γ released in response to antigen challenge is "captured" in the immediate surroundings of the activated cell. After addition of a second tracer monoclonal antibody (mAb) and subsequent color development steps, discrete spots are generated that reflect the number of cytokine-secreting cells. In the second procedure, cells are stimulated for a few hours with antigen prior to intracellular staining with fluorochrome-labeled antibody specific for IFN-γ. Cytokine-producing cells are then counted by microfluorometric flow cytometry.

5.4.2 Enumeration of Antigen-Specific Lymphocytes with Soluble Polymeric and Fluorescent Antigen Ligands

The development of soluble MHC class I and II/peptide complexes that have sufficient affinity/avidity to bind stably to the cognate TCR α/β variable regions has constituted a technological breakthrough for the field of cell-mediated immunity. Indeed, this new type of reagent has now allowed to take a direct look at antigen-specific T cell responses. The main current technique for their preparation takes advantage of the avidin molecule to generate tetrameric antigen ligand complexes (Altman et al. 1996). Alternatively, divalent MHC class I constructs have been generated using Ig as scaffold (Greten et al. 1998). Fluorescent tetrameric complexes have been used to measure mostly antiviral CTL responses (Altman et al. 1996; Callan et al. 1998; Flynn et al. 1998; Gallimore et al. 1998; Kuroda et al. 1998; Murali-Krishna et al. 1998; Ogg et al. 1998a; Sourdive et al. 1998; Wilson et al. 1998). More recently they have also been used to measure specific T cell responses in the course of experimental bacterial infections (Busch et al. 1998). A complete list of the tetramers reported to date is given in Table 3. Of great interest, it is now possible to use tetramers of MHC class II molecules. These will allow to quantitate helper T cell responses directed against tumor antigens.

We reported the first study of ex vivo enumeration of tumor antigen-specific T cells in cancer patients using fluorescent tetramers (Romero et al. 1998b). In this study, HLA-A2 tetramers containing either the tyrosinase 368–376 or the Melan-A antigenic peptide analogue 26–35 A27L were used in conjunction with anti-CD8 and anti-CD3 mAbs to visualize tumor antigen reactive T lymphocytes present in metastatic lymph nodes (Fig. 1). While no tyrosinase reactive $CD8^+$ lymphocytes were detectable, very high numbers of Melan-A reactive $CD8^+$ lymphocytes were present in all of a series of nine metastatic lymph nodes resected from six HLA-A2 melanoma patients (Fig. 2). It is worth noting that the limit of detection was relatively high, 0.25% of $CD8^+$ lymph node cells. This corresponds to frequencies higher than 1 in 400 $CD8^+$ lymphocytes. The levels of Melan-A $tetramer^+$ $CD8^+$ lymph node cells varied between 0.5% and 3.5% which correspond to frequencies of 1 in 200 to 1 in 30. The levels of $tetramer^+$ $CD8^+$ lymph node cells in normal lymph nodes or in micrometastatic lymph nodes

Table 3. List of reported MHC/peptide tetramers[a]

MHC class	Allele	Antigenic peptide source	References
I	A*0201	Influenza, melanoma, HIV, HTLV-1, EBV	Callan et al. (1998); Dunbar et al. (1998); Ogg et al. (1998a); Romero et al. (1998b); Wilson et al. (1998); Bieganowska et al. (1999); Tan et al. (1999); Yee et al. (1999)
	A11	EBV	Tan et al. (1999)
	B8	EBV	Callan et al. (1998)
	B27	HIV	Wilson et al. (1998)
	B*3501	HIV	Ogg et al. (1998a)
	E*0101	Leader peptides	Braud et al. (1998)
	G1	Human histone peptide	Allan et al. (1999)
	Mamu-A*01	SIV	Kuroda et al. (1998)
	D^b	LCMV, herpes, Sendai, influenza, Theiler's virus	Flynn et al. (1998); Gallimore et al. (1998); Murali-Krishna et al. (1998); Stevenson et al. (1998); Johnson et al. (1999)
	D^k	Polyoma virus	Wilson et al. (1999)
	K^b	Mouse herpes virus 68	Stevenson et al. (1998)
	K^d	*Listeria monocytogenes*, HLA-Cw3, -C7	Bousso et al. (1998); Busch et al. (1998); Calbo et al. (1999)
	L^d	LCMV	Murali-Krishna et al. (1998); Sourdive et al. (1998)
	Qa-1^b	Signal peptide	Salcedo et al. (1998); Vance et al. (1998)
II	I-A^d	Ovalbumin	Crawford et al. (1998)
	I-E^k	Mouse cytochrome C, hemoglobin	Crawford et al. (1998); Gütgemann et al. (1998)

LCMV, lymphocytic choriomeningitis virus; SIV, simian immunodeficiency virus
[a]This is a compilation of MHC/peptide tetramers reported until approximately August 1999. All of the class I MHC tetramers were prepared following the same experimental approach described by Altman et al. (1996)

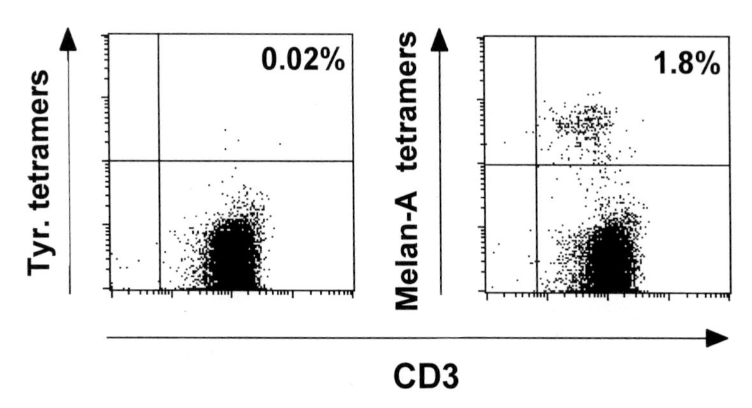

Fig. 1. Visualization of HLA-A2/tumor peptide antigen tetramer binding lymphocytes in tumor infiltrated lymph nodes (TILNs) without prior in vitro culture. A metastatic lymph node from an HLA-A2 melanoma patient was prepared as a single cell suspension and then stained with either A2/Melan-A 26–35 A27L or A2/tyrosinase (Tyr.) 368–376 tetramers together with anti-$CD3^{PerCP}$ and anti-$CD8^{FITC}$. *Dot plots* are shown for gated $CD8^{+}$ lymph node cells. Reproduced with minor modifications from Romero et al. (1998b) by copyright permission of The Rockefeller University Press

were below the detection limit, suggesting a tumor antigen dependent accumulation of $tetramer^{+}$ cells in the metastatic lesions.

The ability to visualize specific T cells in an antigen-dependent manner has several advantages. The major one is that, for the first time in the 30 years since CTLs were first described, it is possible to enumerate these cells without the need to use indirect functional assays. In addition, this new technology enables rapid and clean separation of homogenous populations of antigen-specific T lymphocytes by flow cytometry cell sorting. These populations are a unique source of cells for TCR repertoire analysis and, importantly, for antigen-targeted adoptive transfer therapy (Valmori et al. 1999a). Tetramer staining also allows to perform extensive phenotyping using the large panel of cell surface markers available for human T lymphocytes including markers associated with cell activation status, homing, costimulatory receptors, killer activatory or inhibitory receptors, death receptors, integrins, etc. In addition, intracellular proteins may also be targeted for flow cytometric analyses in conjunction with tetramer staining. These would include

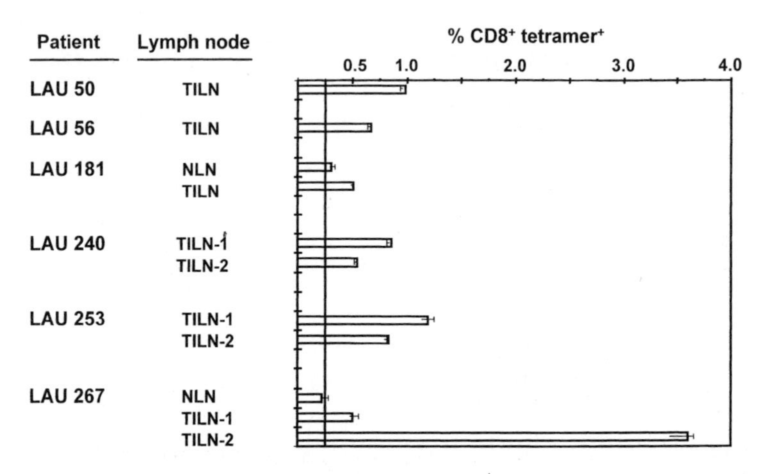

Fig. 2. High frequencies of A2/Melan-A tetramer$^+$ lymphocytes in a series of ex vivo tumor infiltrated lymph nodes (*TILN*) from HLA-A2$^+$ metastatic melanoma patients. Aliquots of TILNs or normal lymph nodes (*NLNs*) were analyzed by three-color flow cytometry after staining with anti-CD3 and anti-CD8 and A2/Melan-A tetramers as in Fig. 1. A cut-off value of 0.25% (as indicated by the *corresponding line*) was calculated as the mean of a total of 22 determinations (0.11) of CD8$^+$ tetramer$^+$ lymph node (LN) cells from 12 LNs analyzed from non-melanoma tumors, plus 3 SD (0.14%). These included five HLA-A2$^+$ and seven HLA-A2$^-$ non-melanoma cancer patients (Romero et al. 1998b). The data represented in the figure are a compilation of part of the results presented in the Table 3 of Romero et al. (1998b). The data used to calculate the cut-off or detection limit were reported in Table 2 of the same reference

kinases and adaptor molecules associated with TCR signaling, cytokines, perforin, granzyme B, and components of the different phases of the cell cycle.

We have initiated the assessment of the differentiation state of Melan-A tetramer$^+$ lymphocytes found in the metastatic lymph nodes ex vivo. As illustrated in Fig. 3, the majority of tetramer$^+$ lymph node cells expressed high levels of the CD45RO marker and low levels of the CD45RA isoform consistent with a memory/activated phenotype. Interestingly, the lower numbers of tetramer$^+$ lymphocytes present in a lymph node resected from the same anatomical region and not infiltrated by tumor had the reciprocal pattern of CD45 isoform expression

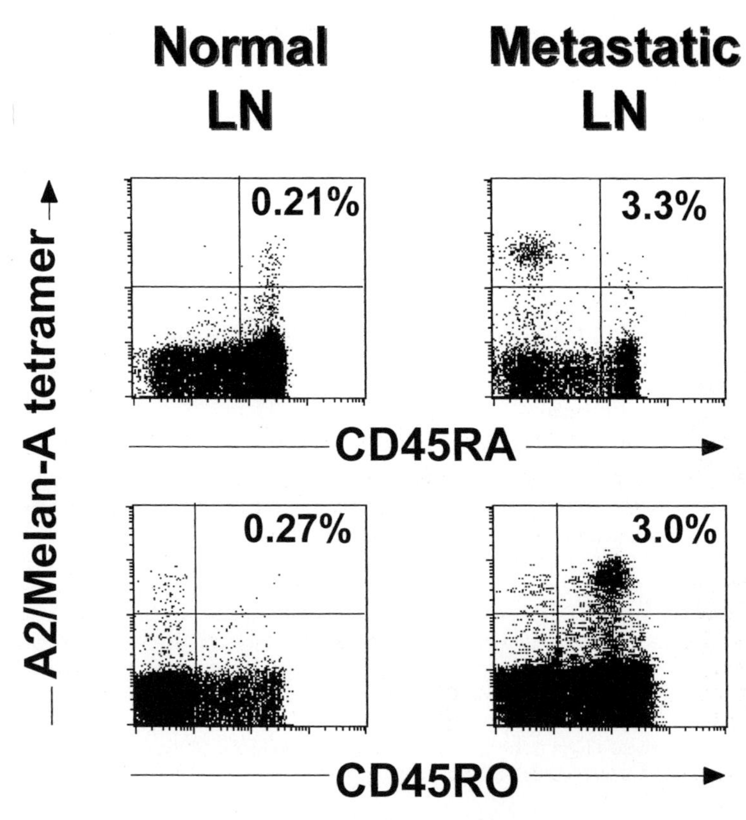

Fig. 3. Ex vivo A2/Melan-A tetramer$^+$ tumor infiltrated lymph nodes (TILNs) have an activated phenotype. Cell suspensions obtained from normal (NLN) or metastatic (TILN) lymph nodes from patient LAU 267 were directly analyzed by three-color flow cytometry using anti-CD8^{PerCP} mAb, A2/Melan-A tetramers, and either anti-CD45^{FITC} (*top panels*) or anti-CD45RO^{FITC} mAb (*bottom*). *Dot plots* are shown for gated CD8$^+$ LN cells. Reproduced with minor modifications from (Romero et al. 1998b) by copyright permission of The Rockefeller University Press

Table 4. Percentages of A2/Melan-A$^+$ and A2/Flu-MA$^+$ cells in circulating CD8$^+$ T cells[a]

	HLA-A2 (–)[b]	HLA-A2 (+)		
	Healthy (*n*=9)	Healthy (*n*=10)	Melanoma (*n*=10)	Melanoma and vitiligo (*n*=3)
Melan-A$^+$				
Mean	0.014	0.07	0.07	0.23
Cut-off[c]	0.04			
Number positive[d]	NA	6/10	7/10	3/3
Flu-MA$^+$				
Mean	0.001	0.12	0.31	0.09
Cut-off[c]	0.01			
Number positive[d]	NA	10/10	8/8	3/3

NA, not applicable
[a]These figures are a compilation of a large flow cytometry study performed using fluorescent HLA-A2/Melan-A and HLA-A2/influenza tetramers on a group of 32 individuals, 9 of them HLA-A2 (–) and the remaining 23 HLA-A2 (+) as described in detail (Pittet et al. (1999)
[b]The mean percentage of tetramer-positive lymphocytes detected in a group of nine HLA-A2 (–) individuals indicated the background tetramer staining level
[c]The cut-off values for each tetramer were determined as the mean plus three standard deviations. Percentages of tetramer$^+$ lymphocytes higher than these cut-off values were considered significant
[d]Number of individuals in each group with significant levels of tetramer$^+$ lymphocytes over the total of individuals analyzed

consistent with a naïve phenotype. Also in contrast with the antigen experienced phenotype in the metastatic lymph node, the majority of Melan-A tetramer$^+$ CD8$^+$ lymphocytes that could be detected in the peripheral blood from the same patient had a naïve phenotype (Romero et al. 1998b). Thus, the marked bias toward the CD45RO$^+$/CD45RA$^-$ memory phenotype in Melan-A tetramer$^+$ lymphocytes in metastatic lymph nodes was specific to the site of tumor contact and indicated that an antigen-specific immune response had been triggered in these patients.

We have also assessed the levels of Melan-A- and tyrosinase-reactive T lymphocytes in the circulating compartment of melanoma patients and of a group of healthy HLA-A2 individuals. Since most adults are sensitized to influenza virus antigens, we also included an HLA-A2 tetramer containing the immunodominant HLA-A2-restricted CTL epitope influenza matrix 58–66 (Lehner et al. 1995). Indeed, we found that all individuals examined had detectable levels of influenza tetramer$^+$ lymphocytes in the circulating lymphocyte pool (Table 4). As expected, the majority of influenza tetramer$^+$ lymphocytes had a memory/activated phenotype. In contrast, and similar to the findings in the lymph nodes, no detectable levels of tyrosinase reactive lymphocytes were apparent in the circulating lymphocyte compartment. However, this did not reflect absence of CTL precursors specific for this melanocyte/melanoma differentiation antigen since it was possible to detect tetramer reactive lymphocytes in 6/10 patients tested following a single in vitro stimulation of their PBL with the tyrosinase synthetic peptide (Valmori et al. 1999b).

This finding illustrates the important issue of sensitivity concerning the use of tetramers to monitor antigen-specific lymphocyte responses ex vivo. Indeed, mainly the background levels for flow cytometry cell counting determine the detection limit of tetramer staining. We determined a detection limit of approximately 0.04% of CD8$^+$ lymphocytes (Pittet et al. 1999). This value is approximately tenfold more sensitive than that determined for lymph node cell suspensions and it allows detection of relatively strong T cell responses directed against a single epitope. Indeed, the value of 0.04% corresponds to a frequency of 1/2500 CD8$^+$ lymphocytes (a frequency of 1/25,000 PBL assuming that CD8$^+$ are in general 10% of PBL). Thus, a negative result in ex vivo tetramer-stained PBL should be interpreted with caution as it means that the frequency of specific lymphocytes is lower than 1/5000 CD8$^+$ lymphocytes.

In marked contrast to the picture obtained with the tyrosinase tetramers, high numbers of circulating Melan-A tetramer$^+$ lymphocytes were readily detectable ex vivo in the majority of cancer patients tested (Table 4). These included 7/10 HLA-A2 melanoma patients with metastatic disease and 3/3 melanoma patients that had concurrent vitiligo (Pittet et al. 1999). The latter observation is intriguing since vitiligo might be caused by autoimmune destruction of melanocytes. In fact, it has been reported that melanoma patients with vitiligo may have an improved prognosis as compared to the group of melanoma patients without vitiligo (Nordlund et al. 1983). Moreover, we found increased numbers of circulating Melan-A tetramer$^+$ lymphocytes expressing the cutaneous lymphocyte antigen (which has been implicated in lymphocyte homing to the skin) in patients with vitiligo but without melanoma (Ogg et al. 1998b).

A surprising finding of these studies was that similar to metastatic melanoma patients, high numbers of circulating Melan-A tetramer$^+$ lymphocytes were also found in 6/10 healthy donors (Table 4). Although the mean level of Melan-A tetramer$^+$ lymphocytes was comparable in the two groups, variable proportions of tetramer$^+$ lymphocytes with an activated/memory phenotype were only present in PBL from some melanoma patients (Pittet et al. 1999). The mechanisms involved in the selection and maintenance of high numbers of circulating Melan-A reactive lymphocytes with a naïve phenotype are poorly understood. They, however, may provide opportunities for Melan-A-based immunotherapy of melanoma.

As mentioned above, ex vivo enumeration of antigen-specific T lymphocytes can also be achieved by IFN-γ ELISPOT. The detection limit for this assay was around 1 in 10^5 recently thawed PBL. This is based on the observation that there are usually less than 10 spots in 10^6 PBL controls stimulated with an irrelevant peptide (Pittet et al. 1999). Thus, from this stand point (the detection limit), the ex vivo IFN-γ ELISPOT assay appears to be one order of magnitude more sensitive than tetramer-based T cell enumeration. We found that the numbers of influenza matrix reactive cells detected in this assay for all healthy donors and melanoma patients were in good agreement with the numbers determined by tetramer staining and flow cytometry analysis. However, in this case the numbers of IFN-γ spot forming cells are consistenty four- to tenfold lower than the numbers of tetramer$^+$ CD8$^+$ circulating lymphocytes (Fig. 4). Thus, the apparent superior sensitivity of the IFN-γ ELISPOT assay is canceled out by the ability of tetramers to detect many more T lymphocytes. It is unclear at the moment whether this is purely due to a sensitivity issue, or whether this reflects that only a fraction of antigen-specific T cells is able to produce IFN-γ.

In contrast to the influenza matrix-specific CTLs, we observed a nearly complete dissociation between ELISPOT IFN-γ and tetramer-based enumeration of Melan-A-specific T lymphocytes. Indeed, the significantly elevated numbers of Melan-A tetramer$^+$ lymphocytes found in both healthy individuals and metastatic melanoma patients

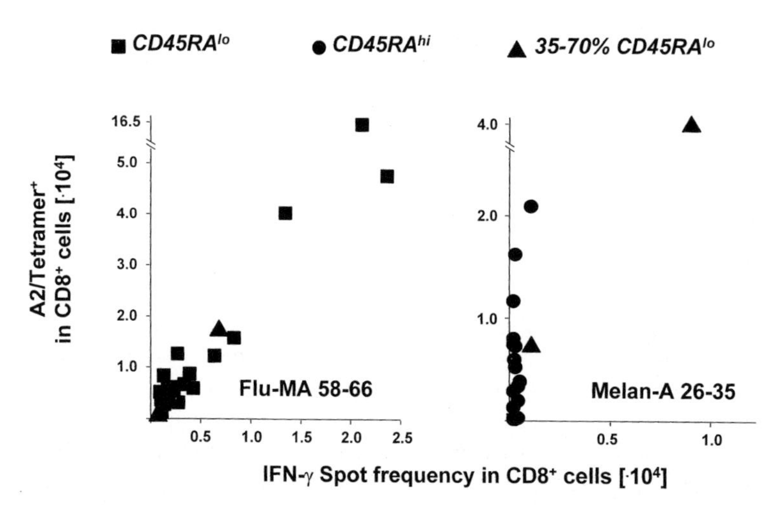

Fig. 4. Relationship between tetramer$^+$ CD8$^+$ cell numbers and ex vivo interferon gamma (IFN-γ) spot forming cell counts. Frequencies of influenza matrix (*left*)- and Melan-A (*right*)-specific CTLs in CD8$^+$ cells from 10 healthy donors and 11 melanoma patients were measured by both IFN-γ ELISPOT assay (*horizontal axis*) and tetramer staining (*vertical axis*). Reproduced with minor modifications from Pittet et al. (1999) by copyright permission of The Rockefeller University Press

could not be detected (Pittet et al. 1999) using the ex vivo IFN-γ ELISPOT assay with two exceptions. These were precisely those patients with a mixture of naïve and memory phenotype Melan-A tetramer$^+$ lymphocytes (Fig. 4). Together these results indicate that the ex vivo ELISPOT assay detects memory type responses and also confirm the naïve cell surface phenotype of circulating Melan-A tetramer$^+$ lymphocytes. It also follows from the discrepancies between tetramer-based and IFN-γ-based T cell counting that functional assays must be accompanied by direct tetramer quantitation.

The analysis of the functional status of tumor reactive T lymphocytes in cancer patients is only at its preliminary phase. It is interesting to note that high levels of tyrosinase reactive lymphocytes were recently reported to display a profound anergic phenotype in the blood of a metastatic melanoma patient (Lee et al. 1999). Although these cells displayed several of the traits of effector T cells, they were found to be functionally unresponsive, unable to directly lyse melanoma target cells or to produce cytokines in response to mitogens, indicating that the clonally expanded tumor antigen-specific T cell population has been selectively rendered anergic in vivo. Whether tumor antigen-specific T cells in the circulating compartment of cancer patients are frequently anergic or tolerant remains controversial. Clearly, many more patients need to be analyzed in detail in order to address this issue.

The underlying molecular events responsible for such a state of unresponsiveness are unclear. We have studied one of such possible mechanisms. This relates to a recently described set of inhibitory receptors known as killer inhibitory receptors or KIRs. While these receptors are expressed in relatively large subpopulations of natural killer (NK) cells, they have also been detected in small proportions of CD8 circulating T lymphocytes in humans (Mingari et al. 1996; Speiser et al. 1998). We have found that some melanoma patients may bear tumor antigen tetramer^{+} lymphocytes expressing NK receptors and that the lysis of melanoma cells by these lymphocytes, in some cases, was inhibited by the NK receptor CD94/NKG2 A (Speiser et al. 1999). Thus, it is possible that CTL activity against tumor cells may be decreased through NK receptor triggering in vivo.

Finally, the use of tetramers to directly visualize antigen-specific T lymphocytes has uncovered an interesting aspect of T cell biology. It has been noted that the kinetics of tetramer binding may differ for T cell populations. Thus, those T cells from a primary immune reaction may display a tetramer dissociation rate faster than that displayed by T cell populations expanding in response to a second encounter with antigen (Busch and Pamer 1999; Savage et al. 1999). This may reflect the selection during the secondary response of T cells expressing TCRs of higher average affinity for peptide/MHC than cells in the primary response. It has also been shown recently that tetramers may be used to selectively identify high avidity tumor reactive CTLs and to enrich from a heterogeneous population the subpopulation of peptide reactive T cells that can efficiently lyse tumor target cells (Yee et al. 1999).

5.5 Conclusions and Future Challenges

Recent developments allow a more quantitative and direct approach to monitoring tumor-specific T cell responses. Both ELISPOT IFN-γ and fluorescent MHC/peptide tetramers are perhaps the most potent and sensitive tools available to enumerate antigen-specific T lymphocytes. In addition, tetramers also permit the characterization of antigen-specific T lymphocytes in terms of cell surface phenotype and differentiation state. The ability to directly stain intact antigen-specific T lymphocytes with tetramers enables their isolation as homogenous, unmanipulated populations. Thus, the analytical power of PCR-based techniques to type TCR [immunoscope or CDR3-spectratyping (Pannetier et al. 1993)] can now be fully exploited. Furthermore, these populations may constitute the optimal sources of tumor reactive lymphocytes for in vitro expansion and adoptive transfer to cancer patients.

It is not long from the time when tetramers may be commercially available as reagents of well-defined molecular composition and potency. Efforts will be required to standardize their use in the context of clinical trials of specific immunotherapy. However, since it might be prohibitively expensive to generate tetramers for every single MHC/peptide combination, it may be that the ELISPOT IFN-γ becomes the assay of choice for the high throughput monitoring of antigen-specific T cell responses. Selected MHC/peptide tetramers may then be very valuable for the in depth analysis of those T cell responses of special interest in various clinical and fundamental research settings.

The results of monitoring of T cell responses to certain tumor-associated antigens in cancer patients have confirmed and extended previous observations indicating the existence of active immune surveillance of tumors. It is becoming increasingly clear that many tumors may be immunogenic and that most cancer patients may in fact mount both humoral and T cell-mediated immune responses against multiple tumor antigen targets. The key question in today's tumor immunology is why these spontaneously arising anti-tumor immune responses eventually fail to effectively counter tumor progression. Issues such as tumor escape (Ferrone and Marincola 1995) and T cell dysfunction (Mizoguchi et al. 1992; Speiser et al. 1997; Lee et al. 1999) are being actively examined. In this context, an area that needs urgent development is that of the tumor-T cell interactions at the tumor sites. It is still unclear the paths followed by T lymphocytes in their transit through various compartments of the immune system and in their trafficking and homing to the tumor stroma. Future immunohistochemical procedures that allow the use of tetramers for the identification in situ of antigen-reactive T lymphocytes will be the next breakthrough needed to help address these questions.

Acknowledgements. We thank M. van Overloop for assistance with manuscript preparation. This work was supported in part by the Leenaards Foundation and M.J.P. by the Swiss Cancer League, grant KFS 633-2-1998. We are also grateful to the melanoma patients for their enthusiastic and generous contribution to the studies reviewed here.

References

Allan DSJ, Colonna M, Lanier LL, Churakova TD, Abrams JS, Ellis SA, McMichael AJ, Braud VM (1999) Tetrameric complexes of human histocompatibility leukocyte antigen (HLA)-G bind to peripheral blood myelomonocytic cells. J Exp Med 189:1149–1155

Altman JD, Moss PAH, Goulder PJR, Barouch DH, McHeyzer-Williams MG, Bell JI, McMichael AJ, Davis MM (1996) Phenotypic analysis of antigen-specific T lymphocytes. Science 274:94–96

Bakker ABH, Burg SH van der, Huijbens RJF, Drijfhout JW, Melief CJM, Adema GJ, Figdor CG (1997) Analogues of CTL epitopes with improved MHC class-I binding capacity elicit anti-melanoma CTL recognizing the wild-type eptitope. Int J Cancer 70:302–309

Bieganowska KD, Höllsberg P, Buckle GJ, Lim DG, Greten TF, Schneck J, Altman JD, Jacobson S, Ledis SL, Hanchard B, Chin J, Morgan O, Roth PA, Hafler DA (1999) Direct analysis of viral-specific CD8+ T cells with soluble HLA-A2/Tax11–19 tetramer complexes in patients with human T cell lymphotropic virus-associated myelopathy. J Immunol 162:1765–1771

Bousso P, Casrouge A, Altman JD, Haury M, Kanellopoulos J, Abastado JP, Kourilsky P (1998) Individual variations in the murine T cell response to a specific peptide reflect variability in naive repertoires. Immunity 9:169–178

Braud VM, Allan DSJ, O'Callaghan CA, Söderström K, D'Andrea A, Ogg GS, Lazetic S, Young NT, Bell JI, Phillips JH, Lanier LL, McMichael AJ (1998) HLA-E binds to natural killer cell receptors CD94/NKG2 A, B and C. Nature 391:795–799

Busch DH, Pamer EG (1999) T cell affinity maturation by selective expansion during infection. J Exp Med 189:701–710

Busch DH, Pilip IM, Vijh S, Pamer EG (1998) Coordinate regulation of complex T cell populations responding to bacterial infection. Immunity 8:353–362

Calbo S, Guichard G, Bousso P, Muller S, Kourilsky P, Briand JP, Abastado JP (1999) Role of peptide backbone in T cell recognition. J Immunol 162:4657–4662

Callan MFC, Tan L, Annels N, Ogg GS, Wilson JDK, O'Callaghan CA, Steven N, McMichael AJ, Rickinson AB (1998) Direct visualization of antigen-specific CD8+ T cells during the primary immune response to Epstein-Barr Virus in vivo. J Exp Med 187:1395–1402

Casanova JL, Cerottini J-C, Matthes M, Necker A, Gournier H, Barra C, Widmann C, MacDonald HR, Lemonnier F, Malissen B, Maryanski JL (1992) H-2-restricted cytolytic T lymphocytes specific for HLA display T cell receptors of limited diversity. J Exp Med 176:439–447

Cole GA, Hogg TL, Woodland DL (1994) The MHC class I-restricted T cell response to Sendai virus infection in C57BL/6 mice: a single immunodominant epitope elicits an extremely diverse repertoire of T cells. Int Immunol 6:1767–1775

Cormier JN, Salgaller ML, Prevette T, Barracchini KC, Rivoltini L, Restifo NP, Rosenberg SA, Marincola FM (1997) Enhancement of cellular immunity in melanoma patients immunized with a peptide from MART-1/Melan-A. Cancer J Sci Am 3:37–44

Crawford F, Kozono H, White J, Marrack P, Kappler J (1998) Detection of antigen-specific T cells with multivalent soluble class II MHC covalent peptide complexes. Immunity 8:675–682

Dunbar PR, Ogg GS, Chen J, Rust N, Bruggen P van der, Cerundolo V (1998) Direct isolation, phenotyping and cloning of low frequency antigen-specific cytotoxic T lymphocytes from peripheral blood. Curr Biol 8:413–416

Ferrone S, Marincola FM (1995) Loss of HLA class I antigens by melanoma cells: molecular mechanisms, functional significance and clinical relevance. Immunol Today 16:487–494

Flynn KJ, Belz GT, Altman KD, Ahmed R, Woodland DL, Doherty PC (1998) Virus-specific CD8+ T cells in primary and secondary influenza pneumonia. Immunity 8:683–691

Gallimore A, Glithero A, Godkin A, Tissot AC, Pluckthun A, Elliott T, Hengartner H, Zinkernagel R (1998) Induction and exhaustion of lymphocytic choriomeningitis virus-specific cytotoxic T lymphocytes visualized using soluble tetrameric major histocompatibility complex class I-peptide complexes. J Exp Med 187:1383–1393

Green DR, Scott DW (1994) Activation-induced apoptosis in lymphocytes. Curr Opin Immunol 6:476–487

Greten TF, Slansky JE, Kubota R, Soldan SS, Jaffee EM, Leist TP, Pardoll DM, Jacobson S, Schneck JP (1998) Direct visualization of antigen-specific T cells: HTLV-1 Tax11–19-specific CD8(+) T cells are activated in peripheral blood and accumulate in cerebrospinal fluid from HAM/TSP patients. Proc Natl Acad Sci USA 95:7568–7573

Gütgemann I, Fahrer AM, Altman JD, Davis MM, Chien Y-H (1998) Induction of rapid T cell activation and tolerance by systemic presentation of an orally administered antigen. Immunity 8:667–673

Jäger E, Ringhoffer M, Dienes HP, Arand M, Karbach J, Jäger D, Ilsemann C, Hagedorn M, Oesch F, Knuth A (1996) Granulocyte-macrophage-colony-stimulating-factor enhances immune responses to melanoma-associated peptides in vivo. Int J Cancer 67:54–62

Johnson AJ, Kariuki-Njenga M, Hansen MJ, Kuhns ST, Chen L, Rodriguez M, Pease LR (1999) Prevalent class I-restricted T-cell response to the Theiler's virus epitope Db:VP2121–130 in the absence of endogenous CD4 help, tumor necrosis factor alpha, gamma interferon, perforin, or costimulation through CD28. J Virol 73:3702–3708

Kuroda MJ, Schmitz JE, Barouch DH, Craiu A, Allen TM, Sette A, Watkins DI, Forman MA, Letvin NL (1998) Analysis of Gag-specific cytotoxic T lymphocytes in simian immunodeficiency virus-infected rhesus monkeys by cell staining with a tetrameric major histocompatibility complex class I-peptide complex. J Exp Med 187:1373–1381

Lee PP, Yee C, Savage PA, Fong L, Brockstedt D, Weber JS, Johnson D, Swetter S, Thompson J, Greenberg PD, Roederer M, Davis MM (1999) Characterization of circulating T cells specific for tumor-associated antigens in melanoma patients. Nat Med 5:677–685

Lehner PJ, Wang ECY, Moss PAH, Williams S, Platt K, Friedman SM, Bell JI, Borysiewicz LK (1995) Human HLA-A0201-restricted cytotoxic T lymphocyte recognition of influenza A is dominated by T cells bearing the Vβ17 gene segment. J Exp Med 181:79–91

Marchand M, Van Baren N, Weynants P, Brichard V, Dréno B, Tessier MH, Rankin E, Parmiani G, Arienti F, Humblet Y, Bourlond A, Vanwijck R, Liénard D, Beauduin M, Dietrich P-Y, Russo V, Kerger J, Masucci G, Jäger E, De Greve J, Atzpodien J, Brasseur F, Coulie PG, Bruggen P van der, Boon T (1999) Tumor regressions observed in patients with metastatic melanoma treated with an antigenic peptide encoded by gene *MAGE-3* and presented by HLA-A1. Int J Cancer 80:219–230

McMichael AJ, O'Callaghan CA (1998) A new look at T cells. J Exp Med 187:1367–1371

Men Y, Miconnet I, Valmori D, Rimoldi D, Cerottini J-C, Romero P (1999) Assessment of immunogenicity of human Melan-A peptide analogues in HLA-A*0201/Kb transgenic mice. J Immunol 162:3566–3573

Mingari MC, Schiavetti F, Ponte M, Vitale C, Maggi E, Romagnani S, Demarest J, Pantaleo G, Fauci AS, Moretta L (1996) Human $CD8^+$ T lymphocyte subsets that express HLA class I-specific inhibitory receptors represent oligoclonally or monoclonally expanded cell populations. Proc Natl Acad Sci USA 93:12433–12438

Mizoguchi H, O'Shea JJ, Longo DL, Loeffler CM, McVicar DW, Ochoa AC (1992) Alterations in signal transduction molecules in T lymphocytes from tumor-bearing mice. Science 258:1795–1798

Murali-Krishna K, Altman JD, Suresh M, Sourdive DJ, Zajac AJ, Miller JD, Slansky J, Ahmed R (1998) Counting antigen-specific CD8 T cells: a reevaluation of bystander activation during viral infection. Immunity 8:177–187

Nestle FO, Alijagic S, Gilliet M, Yuansheng S, Grabbe S, Dummer R, Burg G, Schadendorf D (1998) Vaccination of melanoma patients with peptide- or tumor lysate-pulsed dendritic cells. Nat Med 4:328–332

Nordlund JJ, Kirkwood JM, Forget BM, Milton G, Albert DM, Lerner AB (1983) Vitiligo in patients with metastatic melanoma: a good prognostic sign. J Am Acad Dermatol 9:689–696

Ogg GS, Jin X, Bonhoeffer S, Dunbar PR, Nowak MA, Monard S, Segal JP, Cao Y, Rowland-Jones SL, Cerundolo V, Hurley A, Markowicz M, Ho DD, Nixon DF, McMichael AJ (1998a) Quantitation of HIV-1-specific cytotoxic T lymphocytes and plasma load of viral RNA. Science 279:2103–2106

Ogg GS, Dunbar PR, Romero P, Chen J, Cerundolo V (1998b) High frequency of skin-homing melanocyte-specific cytotoxic T lymphocytes in autoimmune vitiligo. J Exp Med 188:1203–1208

Pannetier C, Cochet M, Darche S, Casrouge A, Zöller M, Kourilsky P (1993) The sizes of the CDR3 hypervariable regions of the murine T-cell receptor β chains vary as a function of the recombined germ-line segments. Proc Natl Acad Sci USA 90:4319–4323

Parkhurst MR, Salgaller ML, Southwood S, Robbins PF, Sette A, Rosenberg SA, Kawakami Y (1996) Improved induction of melanoma-reactive CTL with peptides from the melanoma antigen gp100 modified HLA-A*0201-binding residues. J Immunol 157:2539–2548

Pittet MJ, Valmori D, Dunbar PR, Speiser D, Liénard D, Lejeune F, Fleischhauer K, Cerundolo V, Cerottini J-C, Romero P (1999) High frequencies of naive Melan-A/MART-1-specific CD8+ T cells in a large proportion of HLA-A2 individuals. J Exp Med 190:705–716

Reddish MA, MacLean GD, Koganty RR, Kan-Mitchell J, Jones V, Mitchell MS, Longenecker BM (1998) Anti-MUC1 class I-restricted CTLs in metastatic breast cancer patients immunized with a synthetic MUC1 peptide. Int J Cancer 76:817–823

Romero P, Pannetier C, Herman J, Jongeneel CV, Cerottini J-C, Coulie P (1995) Multiple specificities in the repertoire of a melanoma patient's cytolytic T lymphocytes directed against tumor antigen MAGE-1.A1. J Exp Med 182:1019–1028

Romero P, Gervois N, Schneider J, Escobar P, Valmori D, Pannetier C, Steinle A, Wölfel T, Liénard D, Brichard V, Van Pel A, Jotereau F, Cerottini J-C (1997) Cytolytic T lymphocyte recognition of the immunodominant HLA-A*0201 restricted Melan-A/MART-1 antigenic peptide in melanoma. J Immunol 159:2366–2374

Romero P, Cerottini J-C, Waanders G (1998a) Novel methods to monitor antigen-specific cytotoxic T cell responses in cancer immunotherapy. Mol Med Today 4:305–312

Romero P, Dunbar PR, Valmori D, Pittet MJ, Ogg GS, Rimoldi D, Chen JL, Liénard D, Cerottini J-C, Cerundolo V (1998b) Ex vivo staining of metastatic lymph nodes by class I major histocompatibility complex tetramers reveals high numbers of antigen-experienced tumor-specific cytotoxic T lymphocytes. J Exp Med 188:1641–1650

Rosenberg SA, Yang JC, Schwartzentruber DJ, Hwu P, Marincola FM, Topalian SL, Restifo NP, Dudley ME, Schwarz SL, Spiess PJ, Wunderlich JR, Parkhurst MR, Kawakami Y, Seipp CA, Einhorn JH, White DE (1998) Immunologic and therapeutic evaluation of a synthetic peptide vaccine for the treatment of patients with metastatic melanoma. Nat Med 4:321–327

Salcedo M, Bousso P, Ljunggren HG, Kourilsky P, Abastado JP (1998) The Qa-1b molecule binds to a large subpopulation of murine NK cells. Eur J Immunol 28:4356–4361

Salgaller ML, Marincola FM, Cormier JN, Rosenberg SA (1996) Immunization against epitopes in the human melanoma antigen gp100 following patient immunization with synthetic peptides. Cancer Res 56:4749–4757

Savage PA, Boniface JJ, Davis MM (1999) A kinetic basis for T cell receptor repertoire selection during an immune response. Immunity 10:485–492

Sourdive DJ, Murali-Krishna K, Altman JD, Zajac AJ, Whitmire JK, Pannetier C, Kourilsky P, Evavold B, Sette A, Ahmed R (1998) Conserved T cell receptor repertoire in primary and memory CD8 T cell responses to an acute viral infection. J Exp Med 188:71–82

Speiser DE, Miranda R, Zakarian A, Bachmann MF, McKall-Faienza K, Odermatt B, Hanahan D, Zinkernagel RM, Ohashi PS (1997) Self antigens expressed by solid tumors do not stimulate naive or activated T cells: implications for immunotherapy. J Exp Med 186:645–653

Speiser DE, Valmori D, Rimoldi D, Pittet MJ, Liénard D, MacDonald HR, Cerottini J-C, Romero P (1998) CD28 negative cytotoxic effector T cells frequently express NK receptors and are present at variable proportions in circulating lymphocytes from melanoma patients. Eur J Immunol 29:1900–1999

Speiser DE, Pittet MJ, Valmori D, Dunbar PR, Rimoldi D, Liénard D, MacDonald HR, Cerottini J-C, Cerundolo V, Romero P (1999) In vivo expression of natural killer cell inhibitory receptors by human melanoma-specific cytolytic T lymphocytes. J Exp Med 190:775–782

Stevenson PG, Belz GT, Altman JD, Doherty PC (1998) Virus-specific CD8+ T cell numbers are maintained during γ-herpesvirus reactivation in CD4-deficient mice. Proc Natl Acad Sci USA 95:15565–15570

Tan LC, Gudgeon N, Annels NE, Hansasuta P, O'Callaghan CA, Rowland-Jones S, McMichael AJ, Rickinson AB, Callan MFC (1999) A re-evaluation of the frequency of CD8+ T cells specific for EBV in healthy virus carriers. J Immunol 162:1827–1835

Valmori D, Pittet MJ, Rimoldi D, Liénard D, Dunbar R, Cerundolo V, Lejeune F, Cerottini J-C, Romero P (1999a) An antigen targeted approach to adoptive transfer therapy of cancer. Cancer Res 59:2167–2173

Valmori D, Pittet MJ, Vonarbourg C, Rimoldi D, Liénard D, Speiser D, Dunbar R, Cerundolo V, Cerottini J-C, Romero P (1999b) Analysis of the cytolytic T lymphocyte response of melanoma patients to the naturally HLA-A*0201-associated tyrosinase peptide 368–376. Cancer Res 59:4050–4055

Vance RE, Kraft JR, Altman JD, Jensen PE, Raulet DH (1998) Mouse CD94/NKG2 A is a natural killer cell receptor for the nonclassical major histocompatibility complex (MHC) class I molecule Qa-1. J Exp Med 188:1841–1848

Wilson JDK, Ogg GS, Allen RL, Goulder PJR, Kelleher A, Sewell AK, O'Callaghan CA, Rowland-Jones SL, Callan MFC, McMichael AJ (1998) Oligoclonal expansions of CD8+ T cells in chronic HIV infection are antigen specific. J Exp Med 188:785–790

Wilson CS, Moser JM, Altman JD, Jensen PE, Lukacher AE (1999) Cross-recognition of two middle T protein epitopes by immunodominant polyoma virus-specific CTL. J Immunol 162:3933–3941

Yanagi Y, Tishon A, Lewicki H, Cubitt BA, Oldstone MB (1992) Diversity of T-cell receptors in virus-specific cytotoxic T lymphocytes recognizing three distinct viral epitopes restricted by a single major histocompatibility complex molecule. J Virol 66:2527–2531

Yee C, Savage PA, Lee PP, Davis MM, Greenberg PD (1999) Isolation of high avidity melanoma-reactive CTL from heterogeneous populations using peptide-MHC tetramers. J Immunol 162:2227–2234

6 Retroviral Vectors for Cancer Gene Therapy

M. Collins

6.1 Introduction

More than 300 phase I clinical trials of gene therapy, involving over 2000 patients, have been initiated over the last 5 years: nearly two-thirds of these are for cancer. These trials have tested the feasibility and safety of gene transduction in patients, as well as identifying the problems involved. It is now clear that further research to improve efficiency and specificity of transduction and control of transgene expression will be necessary for more effective cancer gene therapy protocols. This group's contribution to the gene therapy field has been to make a series of improvements in retroviral vectors, resulting in recombinant viruses with higher titer, resistant to human serum, and which are specifically targeted to tumor cells.

6.2 Gene Therapy Approaches for Cancer

The most frequent approach for cancer gene therapy is to introduce genes expressing immunostimulatory molecules, such as cytokines, tumor antigens, and costimulatory molecules. Clinical protocols are based on transduction of autologous tumor, fibroblast, or lymphocyte cell cultures followed by injection of the cells back into the patients. Many tumor vaccine studies, using modified tumor cells to enhance the anti-tumor immune response, have described promising immune stimulation and partial tumor responses. We have recently completed our own clinical trial based on treatment of melanoma patients with interleukin-2-secreting tumor cells (Palmer et al. 1999). However, this strategy requires costly, time-consuming cell culture. In order to assess many candidate immunostimulatory genes for their ability to induce tumor immunity, reliable in vivo gene delivery to tumor cells is required.

Other approaches in cancer gene therapy clinical trials are gene directed enzyme prodrug therapy (GDEPT), tumor suppressor gene replacement, and oncogene inactivation. While tumor regression has been observed following GDEPT in some patients with brain tumors, animal studies have shown that cytokine gene vaccination plus GDEPT, or suppressor gene replacement plus chemotherapy resulted in synergistic anti-tumor effects (Castleden et al. 1997).Tumor vasculature has a critical role in supporting solid tumor expansion and has become the focus of much attention for cancer therapy. Recent experiments using endostatin, an endogenous inhibitor of angiogenesis, show particular promise in treating established tumors in a murine model, with several cycles of treatment resulting in prolonged dormancy (Boehm et al. 1997). Several anti-angiogenic chemotherapeutics are also currently undergoing clinical trial. In order to develop an effective anti-tumor or anti-vascular approach, we are developing vectors for gene therapy targeted to tumors and tumor vasculature (Martin et al. 1999).

6.3 Choice of Vector

Although gene delivery methodology has substantially improved in recent years, no existing method is ideal. The available vectors are either viral, derived from a variety of DNA and RNA viruses, or non-viral, based on delivery of DNA alone or in combination with cationic liposomes. Viral vectors account for nearly 75% of all currently approved gene therapy protocols, approximately 70% of which are using retroviruses: adenoviruses have also found significant application, whilst adeno-associated virus (AAV) is used in relatively few cases. Vectors based on other viruses, such as herpes simplex virus (HSV), are also under development. Liposome-mediated gene delivery is the second most popular vector system and accounts for the majority of non-viral protocols.

Each delivery system has its own merits and problems, depending on the particular application. Vectors based on retroviruses integrate efficiently in host chromosomal DNA, whereas other vectors result in only transient transgene expression. Some DNA virus vectors, including those derived from adenovirus and HSV, are problematic in that the vector itself elicits an immune response in patients due to expression of viral proteins. A further potential complication for clinical application is that prior exposure to adenoviruses may limit gene transfer efficiency due to antiviral antibodies. HSV vectors are most suited for gene transfer to neuronal cells, but are still somewhat cytopathic due to expression of viral genes and may also suffer from pre-existing immunity. AAV vectors may integrate in some target cells, although without the chromosomal targeting which is a feature of the virus itself, but have relatively little capacity (maximum 4.5 kb), preventing the coordinate delivery of two genes for combination therapy.

Of the non-viral gene delivery methods the use of cationic liposomes is the most well developed, with others based on protein/DNA complexes at an early stage of development. The use of naked DNA is peculiar to gene delivery to muscle. Whilst there is essentially no constraint on capacity as with the viral vectors and there are no viral proteins to generate immunogenicity, non-viral DNA delivery is relatively inefficient and expression is transient. For some tumor therapy approaches, transient expression of transgenes may be sufficient and an antiviral response may be advantageous. However, for immunostimulatory approaches, the use of non-immunogenic vectors at the development stage is crucial in order to assess accurately the effects of immunostimulatory transgenes in the absence of background immune response. Additionally, the lack of an anti-vector response is important if repeated administration of the vector (or a similar vector) is required.

Prolonged transgene expression is also likely to be desirable for some therapeutic applications, such as induction of an immune response by cytokine-expressing tumor cells or elimination of tk-positive cells by repeated ganciclovir administration.

6.4 Retroviral Vectors

Helper-free recombinant retroviral vectors are simply generated, stably integrate, and are not immunogenic since they do not encode virus proteins. The disadvantages of conventional retroviral vectors include their rapid inactivation by complement upon exposure to human serum. This precludes many applications which required the transduction of patient cells in situ; in our case we wished to use direct injection of viruses for gene delivery to tumor cells or tumor vasculature. To overcome this problem, we examined the human serum sensitivity of retroviruses produced by a variety of mammalian cells. Human cells produced viruses that survived a 2-h incubation with human serum, whereas viruses produced from animal packaging cell lines were inactivated within 10 min. It was known that humans, apes, and Old World monkeys lack the enzyme for (α1,3)galactosyltransferase [(α1,3)gal]. Because human cells do not express this sugar, we are not tolerant to it and produce high titers of anti-(α1,3)gal antibodies in response to bacteria and other infectious agents. We demonstrated that the inactivation of retroviruses produced by non-human cells is due to triggering of complement by these antibodies in human serum, which react with the retroviral envelope proteins carrying (α1,3)gal epitopes added by the transferases present in the heterologous packaging cells (Takeuchi et al. 1996). We then made packaging cells from human cell lines which produced virus resistant to human serum (Cosset et al. 1995).

Recent developments in packaging cell technology and in vector modification with cationic liposomes have significantly improved titer. We have now been able to transduce human tumor xenografts following intraperitoneal or intratumoral viral injection (Porter et al. 1998; Martin et al. 1999). In terms of clinical safety no problem regarding the use of replication-defective retroviral vectors has been noted to date, despite their widespread use in animal models and clinical trials. However, current adenoviral (and potentially herpesviral) vectors have the problems of generation of a harmful inflammatory response, due to expression of viral proteins, and the possibility of recombination with wild-type viruses present in the recipient.

One major drawback of retroviral vectors based on murine leukemia viruses (MLV) for many gene therapy applications is that they are not able to infect non-dividing cells (Miller et al. 1990). For some applications this may be used to advantage, for example to restrict transduction to proliferating tumor vascular endothelial cells as opposed to quiescent uninvolved cells. However, for gene delivery to tumor cells, only a small proportion of which will be dividing at the time of vector administration, transduction efficiency by MLV-based vectors will definitely be restricted and retroviral vectors without this limitation will be advantageous. Vectors based on human immunodeficiency virus (HIV) are currently under development because lentiviruses, including HIV, are known to proliferate in non-dividing cells, such as macrophages. Using a first generation, transient packaging system, it has been demonstrated that HIV vectors can efficiently transduce non-dividing rat neurons following in vivo administration (Naldini et al. 1996). The development of stable, safe packaging systems based on lentiviruses is a future goal of our laboratory.

References

Boehm T, Folkman J, Browder T, O'Reilly MS (1997) Inhibition of tumour angiogenesis as a strategy to circumvent acquired resistance to anti-cancer therapeutic agents. Nature 390:404–407

Castleden SA, Chong H, Garcia-Ribas I, Melcher AA, Hutchinson G, Roberts B, Hart IR, Vile RG (1997) A family of bicistronic vectors to enhance both local and systemic antitumour effects of HSVtk or cytokine expression in a murine melanoma model. Hum Gene Ther 8:2087–2102

Cosset FL, Takeuchi Y, Battini JL, Weiss RA, Collins MKL (1995) High-titer packaging cells producing recombinant retroviruses resistant to human serum. J Virol 69:7430–7436

Martin F, Neil S, Kupsch J, Maurice M, Cosset F-L, Collins MKL (1999) Retrovirus targeting by tropism restriction to melanoma cells. J Virol 73:6923–6929

Miller DG, Adam MA, Miller AD (1990) Gene transfer by retroviral vectors occurs only in cells that are actively replicating at the time of infection. Mol Cell Biol 10:4239–4242

Naldini L, Blomer U, Gallay P, Ory D, Mulligan R, Gage FH, Verma IM, Trono D (1996) In vivo gene delivery and stable transduction of nondividing cells by a lentiviral vector. Science 272:263–267

Palmer K, Moore J, Everard M, Harris JD, Rodgers S, Rees RC, Murray AK, Mascari R, Kirkwood J, Riches PG, Fisher C, Thomas JM, Johnston SRD, Collins MKL, Gore ME (1999) Gene therapy with autologous, interleukin-2-secreting tumour cells in patients with malignant melanoma increases circulating anti-tumour cytotoxic T lymphocytes. Hum Gene Ther 10:1261–1268

Porter C, Lukacs K, Box G, Takeuchi Y, Collins MKL (1998) Cationic liposomes enhance the rate of transduction by a recombinant retroviral vector in vitro and in vivo. J Virol 72: 4832–4840

Takeuchi Y, Porter CD, Strahan KM, Preece AF, Gustafsson K, Cosset FL, Weiss RA, Collins MKL (1996) Sensitization of cells and retroviruses to human serum by alpha(1–3) galactosyltransferase. Nature 379:85–88

7 CpG Oligonucleotides as Immune Adjuvants

A. M. Krieg

7.1 Introduction

In recent years, it has become increasingly clear that effective vaccination strategies require activation of both the innate and the acquired arms of immune defenses. Vaccines comprising purified protein antigens have been found to induce little or no immune response unless the vaccine also contains components with the ability to activate antigen-presenting cells (APCs). Upon activation, the APCs upregulate their expression of costimulatory molecules such as B7–1 and B7–2, whose expression is essential for the optimal induction of acquired immune responses. Recent studies have demonstrated that many adjuvants have direct stimulatory effects on APCs.

It is now widely accepted that APCs possess pattern recognition receptors (PRRs), which give them a broad ability to detect molecular structures present in many pathogens, but not in host molecules. For example, vertebrates have evolved PRRs to detect microbial structures such as lipopolysaccharide (LPS), high mannose proteins, and viral double-stranded RNA structures (Dempsey et al. 1996; Kumar et al. 1997).

Although DNA has usually been thought of primarily for its function of encoding genetic information, recent studies have also uncovered a structural difference between vertebrate and prokaryotic DNA which allows the detection of the latter by host APCs (Krieg et al. 1995). Tokunaga et al. (1984) were the first to demonstrate the specific immunostimulatory effect of bacterial genomic DNA for activating natural killer (NK) cells and interferon (IFN) secretion (Yamamoto et al. 1988, 1992b). B cell proliferation and immunoglobulin secretion is also specifically stimulated by bacterial but not vertebrate DNA (Messina et al. 1991). It is now clear that these potent immune stimulatory activities of bacterial DNA are due to its content of unmethylated CpG dinucleotides in particular base contexts (Krieg et al. 1995). In contrast to bacterial DNA, in which CpG dinucleotides are unmethylated and are generally present at the expected random frequency of approximately 1/16 bases, it has long been recognized that CpG dinucleotides are usually methylated at the 5 position of the cytosine and "suppressed" in vertebrate genomes, which contain only about 1/4 as many CpG dinucleotides as would be predicted if base utilization was random (Bird 1987). Furthermore, the base context of CpG dinucleotides in vertebrate genomes is not random; CpGs are most frequently preceded by a C and/or followed by a G (Han et al. 1994). If bacterial DNA is methylated with CpG methylase, which converts it into a form more similar to vertebrate DNA, the immune stimulatory activities are lost (Krieg et al. 1995). The immune stimulatory effects of bacterial DNA can be mimicked using synthetic oligodeoxynucleotides (ODN) containing one or more unmethylated CpGs in appropriate base contexts (Table 1). Immune stimulatory CpG ODN can be synthesized using either the native phosphodiester backbone or certain highly nuclease resistant backbones, such as the phosphorothioate backbone, which can greatly improve the immune stimulatory effects by increasing the stability and cellular uptake of the CpG ODN (Zhao et al. 1993, 1996; Krieg et al. 1995, 1996).

Table 1. Identification of optimal stimulatory CpG motifs for murine cytokine production and B cell proliferation. Spleen cells from DBA/2 mice were cultured with the indicated oligodeoxynucleotide (*ODN*) for 24 h, and supernatant IL-12 and IL-10 levels were measured by ELISA using antibodies from Pharmingen (San Diego) as described (Redford et al. 1998). B cell activation was measured by 3H uridine incorporation as described (Krieg et al. 1995). The basal cpm in medium only wells was 524

ODN number	Sequence 5'–3'	IL-12 (pg/ml)	IL-10 (pg/ml)	B cell activation (SI)
Media		<20	108	1
1916	TCCTGACGTTGAAGT	991	1157	6.8
1929	......GC.......	<20	65	1.0
1936	......ZG.......	153	69	1.6
1937	..Z...CG.......	935	377	3.9
1917	.....TCG.......	828	589	4.4
1918	.....GCG.......	776	802	2.8
1919	.....CCG.......	215	684	1.8
1920	....T.CG.......	748	738	4.5
1921	....A.CG.......	772	644	3.7
1922	....C.CG.......	657	616	3.8
1923	......CG A......	571	284	3.5
1924	......CG C......	837	497	3.9
1925	......CGG......	90	168	1.5
1926	......CG.A.....	175	269	3.9
1927	......CG.C.....	332	359	2.4
1928	......CG.G.....	<20	142	2.3
1930	......CG...GG.G	1452	666	1.7
1931	......CG..CCTTC	2031	923	5.4
1935	.....GCGGG.....	<20	43	1.3
1938	....AGCG.......	485	170	7.8
1939	...A..CG.......	872	414	4.4
1940	......CGGG.....	<20	23	0.9
1941	.....GCGG......	<20	34	1.1

7.2 Immune Effects and Mechanisms of CpG DNA

Several immune cell types are directly stimulated by CpG DNA in the absence of any other cells (Fig. 1). Highly purified B cells are stimulated to secrete IL-6, IL-10, and immunoglobulin (Yi et al. 1996b; Redford 1998) and to proliferate in a T cell-independent manner (Krieg et al. 1995). In fact, CpG DNA is a stronger mitogen for B cell proliferation than any other single agent, and can drive more than 95% of B cells into the cell cycle. However, the stimulatory effects of CpG DNA on B cell activation synergize strongly with activation through the antigen receptor, thereby providing a mechanism through which the PRRs of the innate immune system can enhance the generation and amplification of B cell responses (Krieg et al. 1995; Yi et al. 1996b). Another mechanism through which CpG DNA can promote B cell responses is by inhibiting their apoptosis (Yi et al. 1996a, 1998b; Wang et al. 1997; Yi and Krieg 1998a). CpG DNA also acts directly on B cells to make them competent for undergoing isotype switching to either Th1-like or Th2-like isotypes in the appropriate cytokine milieu (Davis et al. 1998). B cells activated by CpG DNA expressed increased surface levels of class II MHC as well as the costimulatory molecules B7–1 and B7–2, suggesting that they should have an improved ability to present antigens to T cells (Krieg et al. 1995; Davis et al. 1998).

CpG DNA is also a potent activator of purified dendritic cell (DC) expression of costimulatory molecules and enhances their ability to drive allogeneic T cell responses (Jakob et al. 1998; Sparwasser et al. 1998; Hartmann et al. 1999). The cytokine production by CpG-activated DCs is dominated by Th1-like cytokines such as IL-12 and IL-18, which strongly promote the development of Th1-like immune responses (Roman et al. 1997; Jakob et al. 1998). CpG DNA also directly activates macrophages to secrete IL-12, TNF-α, and other cytokines (Stacey et al. 1996; Anitescu et al. 1997; Chace et al. 1997). However, the effects of CpG DNA on macrophage antigen processing and presentation seem to be less positive than those exerted on DCs. In fact, macrophages activated by CpG DNA undergo downregulation of surface class II MHC expression and have a reduced ability to present antigens (Chu et al. 1999). CpG DNA activates monocytes to produce IL-6 and TNF-α, although it is not yet clear whether this is a direct or indirect effect (Hartmann and Krieg 1999). The effect of CpG DNA on human mono-

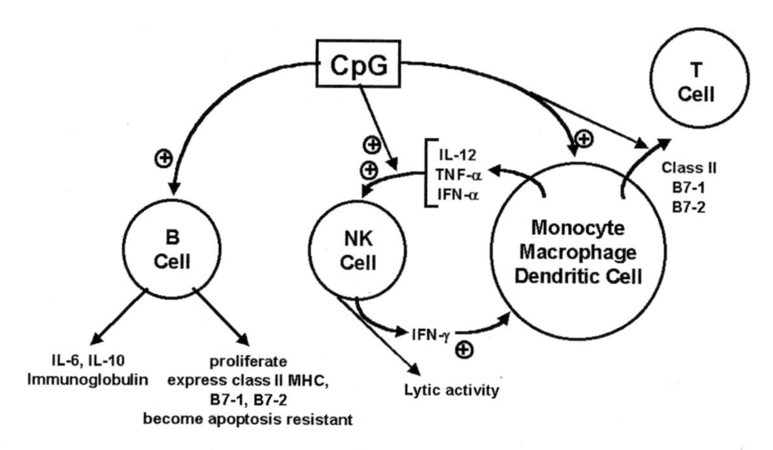

Fig. 1. Cellular immune stimulation by CpG DNA. CpG DNA directly activates B cells to proliferate, express costimulatory molecules, and secrete IL-6, IL-10, and immunoglobulin. CpG DNA also activates monocytes, macrophages, and dendritic cells to secrete Th1-like cytokines and upregulate costimulatory molecule expression. The Th1-like cytokines produced by these antigen-presenting cells (APCs) costimulate natural killer (NK) cells together with CpG DNA to have enhanced interferon (IFN)-γ secretion and lytic activity. T cells are costimulated by presentation of antigens to the T cell receptor by CpG-induced APCS

cytes may be indirect since it is much slower than either the stimulatory effect of LPS or the effect of CpG DNA on other cell types (Hartmann and Krieg 1999).

Several other types of leukocytes are activated by CpG DNA in a costimulatory fashion. For example, NK cells are strongly activated by CpG DNA to secrete IFN-γ and to have increased lytic activity (Yamamoto et al. 1992a; Ballas et al. 1996). However, highly purified NK cells cannot be stimulated by CpG DNA alone. Instead, CpG DNA appears to act together with APC-derived Th1-like cytokines to costimulate NK cell activation (Fig. 1; Ballas et al. 1996; Cowdery et al. 1996; Chace et al. 1997). Like NK cells, CpG DNA does not appear to have direct stimulatory effects on T cells (Lipford et al. 1997; Sun et al. 1998). However, highly purified T cells activated with ligation of the T cell receptor show very strong costimulatory responses to CpG DNA (Bendigs et al. 1999). This costimulatory response is independent of

CD28 (Bendigs et al. 1999). On the other hand, in mixed cell populations, CpG DNA has been reported to inhibit the T cell proliferative response to T cell receptor ligation as a result of the suppressive effects of CpG-induced type I IFNs (Sun et al. 1998). Further studies will be required to define the role of CpG DNA in T cell costimulation in vivo.

The molecular mechanism through which immune cells detect CpG DNA does not appear to rely upon a cell surface receptor, but instead appears to be mediated through one or more intracellular CpG DNA binding proteins (Krieg et al. 1995, manuscript in preparation). DNA of any sequence and methylation status appears equally able to bind to cell surfaces and to be internalized via endocytosis into an endosomal compartment where the DNA is acidified and digested by nucleases (Bennett et al. 1985; Tonkinson and Stein 1994). DNA containing nuclease resistant backbones such as the phosphorothioate backbone have greatly improved resistance to this nuclease degradation and may also have enhanced cellular uptake (Zhao et al. 1993, 1996; Krieg et al. 1996). Endosomal acidification of CpG DNA appears to be required for its immune stimulatory activities since inhibition with specific inhibitors such as chloroquine completely blocks the signaling pathways induced by CpG, but not those induced by LPS, phorbol myristate acetate, or other leukocyte mitogens (MacFarlane and Manzel 1998; Yi et al. 1998a).

Multiple intracellular signaling pathways are rapidly induced in leukocytes exposed to CpG DNA. Both B cells and monocytic cells have increased levels of intracellular reactive oxygen species (Yi et al. 1996b), activation of NFkB (Stacey et al. 1996; Sparwasser et al. 1997b; Yi and Krieg 1998a), and the activation of several mitogen activated protein kinases (Hacker et al. 1998; Yi and Krieg 1998b).

7.3 Therapeutic Applications of CpG DNA in Vaccination

The utility of CpG DNA as a vaccine adjuvant is suggested by several observations. First of all, although any leukocyte mitogen could in principle be suspected of acting as an adjuvant, the fact that CpG DNA synergizes so strongly with B cell activation through the antigen receptor suggests that it should not simply act as a polyclonal mitogen, but rather should specifically enhance immune responses (Krieg et al.

1995). In theory, B cells specific for a microbial antigen would be preferentially activated compared to B cells specific for other antigens in lymphoid tissues exposed to CpG DNA. The synergy between CpG DNA and the B cell antigen receptor appears to be mediated at a very early step in the signaling pathways since it is already seen at the level of activation of the mitogen activated protein kinases (Yi and Krieg manuscript in preparation). Secondly, the enhanced expression of costimulatory molecules and class II MHC molecules on B cells and DCs should enhance their ability to drive T cell responses. Thirdly, the creation of a Th1 cytokine-like milieu in lymphoid tissues exposed to CpG DNA would stimulate the development of Th1-like immune responses, which are generally preferred in therapeutic vaccination. Finally, the ability of CpG DNA to costimulate T cells may further enhance its ability to drive T cell responses.

Indeed, the utility of CpG DNA as a vaccine adjuvant has been demonstrated in numerous model systems. CpG DNA is a strong adjuvant for antibody and Th1-like T cell responses when used as an adjuvant for hen egg lysozyme (Chu et al. 1997), ovalbumin (Lipford et al. 1997), heterologous gammaglobulin (Sun et al. 1997)[42], and β-galactosidase (Roman et al. 1997). In fact, in these systems CpG proved to be a stronger Th1-like adjuvant than complete Freund's adjuvant, despite the fact that the local inflammatory toxic reaction so characteristic of immunization with Freund's was not seen in mice immunized with CpG. CpG DNA is also a potent vaccine adjuvant for infectious disease antigens, including hepatitis B (Davis et al. 1998) and influenza (Moldoveanu et al. 1998). CpG DNA has also proved useful in primate immunization and has been demonstrated to enhance antibody responses to malaria circumsporozooite peptides in Aotus monkeys (Jones et al. 1999), and to the hepatitis B surface antigen in orangutans (Davis et al. manuscript submitted). Several studies have documented utility for CpG DNA as a mucosal adjuvant via the intranasal route (McCluskie and Davis 1998; Moldoveanu et al. 1998). CpG DNA is also effective as an adjuvant for neonatal immunization, where it can overcome poor immune responsiveness (Brazolot Millan et al. 1998; Kovarik et al. 1999).

Therapeutic vaccines are especially important in the field of cancer immunotherapy. In this case, CpG DNA has been shown to be an effective adjuvant for vaccination against B cell lymphoma by inducing protective immunity against lethal tumor challenge (Sun et al. 1996; Weiner et al. 1997). Furthermore, CpG DNA shows strong synergy in tumor vaccination with GMCSF, converting the Th2-like anti-tumor immune response which is seen following immunization with GMCSF into a more Th1-like response with greatly improved survival in therapeutic immunization against lymphoma (Liu et al. 1998). Human clinical trials using CpG DNA as a tumor vaccine adjuvant are expected to begin in 1999.

A final category of therapeutic immunization in which CpG DNA may be expected to play a useful role is that of vaccination against allergic diseases. Allergies are characterized by Th2-like immune responses against harmless environmental antigens. CpG DNA has shown a remarkable ability to effectively reprogram existing Th2-like responses to allergens into Th1-like responses. This has resulted in dramatic inhibition of inflammatory responses to allergen inhalation in mice (Broide et al. 1998; Kline et al. 1998). So far, no adverse effects from allergen inhalation have been observed in mice immunized to have Th1 responses to the allergens.

Overall, these vaccination studies with CpG DNA have shown it to function as an effective adjuvant when simply mixed with any kind of antigen (including live virus vaccines) and administered via almost any route, including intranasal, subcutaneous, intradermal, intramuscular, intravenous, and intraperitoneal. The adjuvant effects of CpG DNA have shown strong synergistic effects with certain other adjuvants. In particular, CpG DNA appears to be most effective when coadministered with a "depot"-like adjuvant such as alum or mineral oil, which may provide more prolonged exposure to antigen and/or CpG DNA. CpG DNA also has been reported to show strong synergy with certain other immune stimulatory adjuvants, most notably GMCSF and QS21.

7.4 Potential Toxicities from Immunization with CpG DNA

Because of the extraordinary power of the immune system, its activation for therapeutic purposes must be regarded with some caution. Overactivation of the immune system is recognized to trigger the systemic inflammatory response syndrome (SIRS), which is associated with high morbidity and mortality. Indeed, treatment with CpG DNA can trigger

SIRS upon endotoxin exposure or in mice sensitized to TNF-α by treatment with d-galactosamine (Sparwasser et al. 1997a). Another potential toxicity of CpG DNA would be the induction of autoimmune diseases. Since CpG DNA is, after all, both a DNA molecule and a potent mitogen, it may be expected that it should activate anti-DNA-specific B cells, leading to the production of high affinity anti-DNA antibodies. Indeed, we had previously hypothesized that this would occur (Krieg 1995). While the possibility cannot be excluded that CpG DNA may under some conditions autoimmunize against DNA, careful studies in both normal and lupus-prone mouse strains have demonstrated no increased induction of autoimmunity by exposure to CpG DNA (Mor et al. 1997). If CpG DNA is conjugated to a foreign protein and administered together with Freund's adjuvant, then the production of anti-DNA antibodies can be elicited (Gilkeson et al. 1996, 1998). However, the severity of lupus in lupus-prone mice immunized with CpG DNA is actually reduced compared to that seen in the lupus-prone mice not exposed to CpG DNA! Overall, the toxicity studies performed with CpG DNA appear to show a wide therapeutic index and provide encouraging evidence that therapeutic immunization with CpG DNA should be safe at effective doses.

7.5 Conclusions

Extensive preclinical studies suggest that CpG DNA is a remarkably effective and safe vaccine adjuvant for prophylactic therapeutic vaccination. Several properties may make CpG ODN preferable to other potential adjuvants. Firstly, ODN are quite easy to synthesize under GMP conditions in large quantities for an economical price. Secondly, ODN are extremely stable to temperature and formulation (as long as the pH is not acidic) and have a shelf life of several years. Thirdly, CpG ODN appear to be effective adjuvants through most if not all routes of immunization and for any kind of vaccine antigen. Finally, preclinical studies suggest that vaccination with CpG ODN will be extremely safe as well as effective. Human clinical trials using CpG ODN are currently underway. Within the next few years, the role of this exciting new immunomodulator in human vaccination will become clear.

Acknowledgements. The author thanks Tilese Arrington and Vickie Akers for secretarial assistance. Financial support was provided through a Career Development Award from the Department of Veterans Affairs and grants from the National Institutes of Health and Coley Pharmaceutical Group.

References

Anitescu M, Chace JH, Tuetken R, Yi A-K, Berg DJ, Krieg AM, Cowdery JS (1997) Interleukin-10 functions in vitro and in vivo to inhibit bacterial DNA-induced secretion of interleukin-12. J Interferon Cytokine Res 17:781

Ballas ZK, Rasmussen WL, Krieg AM (1996) Induction of natural killer activity in murine and human cells by CpG motifs in oligodeoxynucleotides and bacterial DNA. J Immunol 157:1840

Bendigs S, Salzer U, Lipford GB, Wagner H, Heeg K (1999) CpG-oligodeoxynucleotides costimulate primary T cells in the absence of APC. Eur J Immunol 29:1209

Bennett RM, Gabor GT, Merritt MM (1985) DNA binding to human leukocytes. Evidence for a receptor-mediated association, internalization, and degradation of DNA. J Clin Invest 76:2182

Bird AP (1987) CpG islands as gene markers in the vertebrate nucleus. Trends Genet 3:342

Brazolot Millan CL, Weeratna R, Krieg AM, Siegrist CA, Davis HL (1998) CpG DNA can induce strong Th1 humoral and cell-mediated immune responses against hepatitis B surface antigen in young mice. Proc Natl Acad Sci USA 95:15553

Broide D, Schwarze J, Tighe H, Gifford T, Nguyen M-D, Malek S, Van Uden J, Martin-Orozco E, Gelfand EW, Raz E (1998) Immunostimulatory DNA sequences inhibit IL-5, eosinophilic inflammation, and airway hyperresponsiveness in mice. J Immunol 161:7054

Chace JH, Hooker NA, Mildenstein KL, Krieg AM, Cowdery JS (1997) Bacterial DNA-induced NK cell IFN-γ production is dependent on macrophage secretion of IL-12. Clin Immunol Immunopathol 84:185

Chu RS, Targoni OS, Krieg AM, Lehmann PV, Harding CV (1997) CpG oligodeoxynucleotides act as adjuvants that switch on Th1 immunity. J Exp Med 186:1623

Chu RS, Askew D, Noss EH, Tobian A, Krieg AM, Harding CV (1999) CpG oligodeoxynucleotides downregulate macrophage class II MHC antigen processing. J Immunol 163:1188

Cowdery JS, Chace JH, Yi A-K, Krieg AM (1996) Bacterial DNA induces NK cells to produce interferon-γ in vivo and increases the toxicity of lipopolysaccharides. J Immunol 156:4570

Davis HL, Weeratna R, Waldschmidt TJ, Tygrett L, Schorr J, Krieg AM (1998) CpG DNA is a potent adjuvant in mice immunized with recombinant hepatitis B surface antigen. J Immunol 160:870

Dempsey PW, Allison ME, Akkaraju S, Goodnow CC, Fearon DT (1996) C3d of complement as a molecular adjuvant: bridging innate and acquired immunity. Science 271:348

Gilkeson GS, Ruiz P, Pippen AMM, Alexander AL, Lefkowith JB, Pisetsky DS (1996) Modulation of renal disease in autoimmune NZB/NZW mice by immunization with bacterial DNA. J Exp Med 183:1389

Gilkeson GS, Conover J, Halpern M, Pisetsky DS, Feagin A, Klinman DM (1998) Effects of bacterial DNA on cytokine production by (NZB/NZW)F1 mice. J Immunol 161:3890

Hacker H, Mischak H, Miethke T, Liptay S, Schmid R, Sparwasser T, Heeg K, Lipford GB, Wagner H (1998) CpG-DNA-specific activation of antigen-presenting cells requires stress kinase activity and is preceded by non-specific endocytosis and endosomal maturation. EMBO J 17:6230

Han J, Zhu Z, Hsu C, Finley WH (1994) Selection of antisense oligonucleotides on the basis of genomic frequency of the target sequence. Antisense Nucleic Acid Drug Dev 4:53

Hartmann G, Krieg AM (1999) CpG DNA and LPS induce distinct patterns of activation in human monocytes. Gene Ther 6:893

Hartmann G, Weiner G, Krieg AM (1999) CpG DNA as a signal for growth, activation and maturation of human dendritic cells. Proc Natl Acad Sci USA 9305

Jakob T, Walker PS, Krieg AM, Udey MC, Vogel JC (1998) Activation of cutaneous dendritic cells by CpG-containing oligodeoxynucleotides: a role for dendritic cells in the augmentation of Th1 responses by immunostimulatory DNA. J Immunol 161:3042

Jones TR, Obaldia N III, Gramzinski RA, Charoenvit Y, Kolodny N, Davis HL, Krieg AM, Hoffman SL (1999) Synthetic oligodeoxynucleotides containing CpG motifs enhance immunogenicity of a peptide malaria vaccine in Aotus monkeys. Vaccines 17:3065

Kline JN, Waldschmidt TJ, Businga TR, Lemish JE, Weinstock JV, Thorne PS, Krieg AM (1998) Modulation of airway inflammation by CpG oligodeoxynucleotides in a murine model of asthma. J Immunol Cutting Edge 160:2555

Kovarik J, Bozzotti P, Love-Homan L, Pihlgren M, Davis HL, Lambert P-H, Krieg AM, Siegrist C-A (1999) CpG oligodeoxynucleotides can circumvent the Th2 polarization of neonatal responses to vaccines but may fail to fully redirect Th2 responses established by neonatal priming. J Immunol 162:1611

Krieg AM (1995) CpG DNA: a pathogenic factor in systemic lupus erythematosus? J Clin Immunol 15:284

Krieg AM, Yi A-K, Matson S, Waldschmidt TJ, Bishop GA, Teasdale R, Koretzky G, Klinman D (1995) CpG motifs in bacterial DNA trigger direct B-cell activation. Nature 374:546

Krieg AM, Matson S, Herrera C, Fisher E (1996) Oligodeoxynucleotide modifications determine the magnitude of immune stimulation by CpG motifs. Antisense Nucleic Acid Drug Dev 6:133

Kumar A, Yang YL, Flati V, Der S, Kadereit S, Deb A, Haque J, Reis L, Weissmann C, Williams BR (1997) Deficient cytokine signaling in mouse embryo fibroblasts with a targeted deletion in the PKR gene: role of IRF-1 and NF-kB. EMBO J 16:406

Lipford GB, Bauer M, Blank C, Reiter R, Wagner H, Heeg K (1997) CpG-containing synthetic oligonucleotides promote B and cytotoxic T cell responses to protein antigen: a new class of vaccine adjuvants. Eur J Immunol 27:2340

Liu H-M, Newbrough SE, Bhatia SK, Dahle CE, Krieg AM, Weiner GJ (1998) Immunostimulatory CpG oligodeoxynucleotides enhance the immune response to vaccine strategies involving granulocyte-macrophage colony-stimulating factor. Blood 92:3730

MacFarlane DE, Manzel L (1998) Antagonism of immunostimulatory CpG-oligodeoxynucleotides by quinacrine, chloroquine, and structurally related compounds. J Immunol 160:1122

McCluskie MJ, Davis HL (1998) CpG DNA is a potent enhancer of systemic and mucosal immune responses against hepatitis B surface antigen with intranasal administration to mice. J Immunol 161:4463

Messina JP, Gilkeson GS, Pisetsky DS (1991) Stimulation of in vitro murine lymphocyte proliferation by bacterial DNA. J Immunol 147:1759

Moldoveanu Z, Love-Homan L, Huang WQ, Krieg AM (1998) CpG DNA, a novel adjuvant for systemic and mucosal immunization with influenza virus. Vaccine 16:1216

Mor G, Singla M, Steinberg AD, Hoffman SL, Okuda K, Klinman DM (1997) Do DNA vaccines induce autoimmune disease? Hum Gene Ther 8:293

Redford TW, Yi A-K, Ward CT, Krieg AM (1998) Cyclosporine A enhances IL-12 production by CpG motifs in bacterial DNA and synthetic oligodeoxynucleotides. J Immunol 161:3930

Roman M, Martin-Orozco E, Goodman JS, Nguyen M-D, Sato Y, Ronaghy A, Kornbluth RS, Richman DD, Carson DA, Raz E (1997) Immunostimulatory DNA sequences function as T helper-1-promoting adjuvants. Nat Med 3:849

Sparwasser T, Miethke T, Lipford G, Borschert K, Hacker H, Heet K, Wagner H (1997a) Bacterial DNA causes septic shock. Nature 386:336

Sparwasser T, Miethe T, Lipford G, Erdmann A, Hacker H, Heeg K, Wagner H (1997b) Macrophages sense pathogens via DNA motifs: induction of tumor necrosis factor-α-mediated shock. Eur J Immunol 27:1671

Sparwasser T, Koch E-S, Vabulas RM, Heeg K, Lipford GB, Ellwart J, Wagner H (1998) Bacterial DNA and immunostimulatory CpG oligonucleotides trigger maturation and activation of murine dendritic cells. Eur J Immunol 28:2045

Stacey KJ, Sweet MJ, Hume DA (1996) Macrophages ingest and are activated by bacterial DNA. J Immunol 157:2116

Sun S, Cai Z, Langlade-Demoyen P, Kosaka H, Brunmark A, Jackson MR, Peterson PA, Sprent J (1996) Dual function of drosophilia cells as APCs for naive CD8+ T cells: implications for tumor immunotherapy. Immunity 4:555

Sun S, Beard C, Jaenisch R, Jones P, Sprent J (1997) Mitogenicity of DNA from different organisms for murine B cells. J Immunol 159:3119

Sun S, Zhang X, Tough DF, Sprent J (1998) Type I interferon-mediated stimulation of T cells by CpG DNA. J Exp Med 188:2335

Tokunaga T, Yamamoto H, Shimada S, Abe H, Fukuda T, Fujisawa Y, Furutani Y, Yano O, Kataoka T, Sudo T, Makiguchi N, Suganuma T (1984) Antitumor activity of deoxyribonucleic acid fraction from *Mycobacterium bovis* GCG. Isolation, physicochemical characterization, and antitumor activity. J Natl Cancer Inst 72:955

Tonkinson JL, Stein CA (1994) Patterns of intracellular compartmentalization, trafficking and acidification of 5¢-fluorescein labeled phosphodiester and phosphorothioate oligodeoxynucleotides in HL60 cells. Nucl Acids Res 22:4268

Wang Z, Karras JG, Colarusso TP, Foote LC, Rothstein TL (1997) Unmethylated CpG motifs protect murine B lymphocytes against Fas-mediated apoptosis. Cell Immunol 180:162

Weiner GJ, Liu H-M, Wooldridge JE, Dahle CE, Krieg AM (1997) Immunostimulatory oligodeoxynucleotides containing the CpG motif are effective as immune adjuvants in tumor antigen immunization. Proc Natl Acad Sci USA 94:10833

Yamamoto S, Kuramoto E, Shimada S, Tokunaga T (1988) In vitro augmentation of natural killer cell activity and production of interferon-α/β and -γ with deoxyribonucleic acid fraction from *Mycobacterium bovis* BCG. Jpn J Cancer Res 79:866

Yamamoto S, Yamamoto T, Kataoka T, Kuramoto E, Yano O, Tokunaga T (1992a) Unique palindromic sequences in synthetic oligonucleotides are required to induce INF and augment INF-mediated natural killer activity. J Immunol 148:4072

Yamamoto S, Yamamoto T, Shimada S, Kuramoto E, Yano O, Kataoka T, Tokunaga T (1992b) DNA from bacteria, but not from vertebrates, induces interferons, activates natural killer cells and inhibits tumor growth. Microbiol Immunol 36:983

Yi A-K, Krieg AM (1998a) CpG DNA rescue from anti-IgM induced WEHI-231 B lymphoma apoptosis via modulation of IkBa and IkBb and sustained activation of nuclear factor-kB/c-Rel. J Immunol 160:1240

Yi A-K, Krieg AM (1998b) Rapid induction of mitogen activated protein kinases by immune stimulatory CpG DNA. J Immunol 161:4493

Yi A-K, Hornbeck P, Lafrenz DE, Krieg AM (1996a) CpG DNA rescue of murine B lymphoma cells from anti-IgM induced growth arrest and programmed cell death is associated with increased expression of c-myc and bcl-xL. J Immunol 157:4918

Yi A-K, Klinman DM, Martin TL, Matson S, Krieg AM (1996b) Rapid immune activation by CpG motifs in bacterial DNA: systemic induction of IL-6 transcription through an antioxidant-sensitive pathway. J Immunol 157:5394

Yi A-K, Tuetken R, Redford T, Kirsch J, Krieg AM (1998a) CpG motifs in bacterial DNA activates leukocytes through the pH-dependent generation of reactive oxygen species. J Immunol 160:4755

Yi A-K, Chang M, Peckham DW, Krieg AM, Ashman RF (1998b) CpG oligodeoxyribonucleotides rescue mature spleen B cells from spontaneous apoptosis and promote cell cycle entry. J Immunol 160:5898

Zhao Q, Matson S, Herrara CJ, Fisher E, Yu H, Waggoner A, Krieg AM (1993) Comparison of cellular binding and uptake of antisense phosphodiester, phosphorothioate, and mixed phosphorothioate and methylphosphonate oligonucleotides. Antisense Nucleic Acid Drug Dev 3:53

Zhao Q, Temsamani J, Iadarola PL, Jiang Z, Agrawal S (1996) Effect of different chemically modified oligodeoxynucleotides on immune stimulation. Biochem Pharmacol 51:173

8 DNA Vaccination Against Cancer Antigens

F.K. Stevenson, D. Zhu, M.B. Spellerberg, J. Rice, C.A. King, A.R. Thompsett, S.S. Sahota, T.J. Hamblin

8.1 Introduction

Rational approaches aimed at manipulating the immune system to act against cancer cells are now becoming feasible. There are two main reasons for this, both arising from developments in molecular genetics: first, there is a greater understanding of the tumor-associated changes occurring in cancer cells, some of which will generate candidate target antigens. Second, there is increasing knowledge of the processes involved in inducing an effective immune response. In attempting to activate the immune system against cancer, we need to remember that the major task of the immune response is to control or eliminate pathogens. Development of vaccines against cancer will rely on lessons from infectious diseases, and is bringing back together the fields of microbiology and immunology which had become separated, perhaps partly due to the effectiveness of antibiotics.

In terms of impact on disease, vaccination against infectious organisms probably represents the most effective intervention arising from medical research. Following the first successful vaccination against smallpox, which led to complete elimination of this virus, many other infectious diseases can now be effectively prevented. However, it took 200 years to eliminate smallpox, and there are still some infections for which there are no effective vaccines. New strategies for delivering antigens to the immune system are therefore of great interest in the field of infectious diseases. For certain cancers, there is an clear link with infectious organisms. For example, cervical cancer and hepatoma are both associated with viral infection, and successful vaccination against papilloma virus or hepatitis B virus, respectively, should have beneficial preventative effects against these cancers.

We have focused on B cell malignancies, where there is also at least one cancer-associated virus, the Epstein Barr virus (EBV). EBV has a role in inducing lymphomas, shown most clearly in patients undergoing immunosuppressive therapy following allogeneic transplantation. However, there are also several non-viral tumor-associated antigens expressed by B cell tumors (Fig. 1) including mutated proto-oncogene products, such as p53 (Nigro et al. 1989), and products of chromosomal translocations (Rabbitts 1994). Multiple myeloma cells can carry a wealth of candidate antigens, including mutant N-Ras, abnormally glycosylated mucins such as MUC-1 (Takahashi et al. 1994), and the MAGE cancer-testis antigens (Van Baren et al. 1999). Idiotypic determinants, expressed by the clonal Ig of B cells, represent defined sequences unique to each tumor, and are therefore ideal tumor-specific antigens (George and Stevenson 1989). It is known that anti-idiotypic (Id) antibodies can directly mediate death of neoplastic B cells (Racila et al. 1996), and protect against lymphoma (George et al. 1987; Kaminiski et al. 1987). However, idiotypic vaccines have to be made individually, and this technically demanding requirement has been a driving force in the development of DNA vaccines.

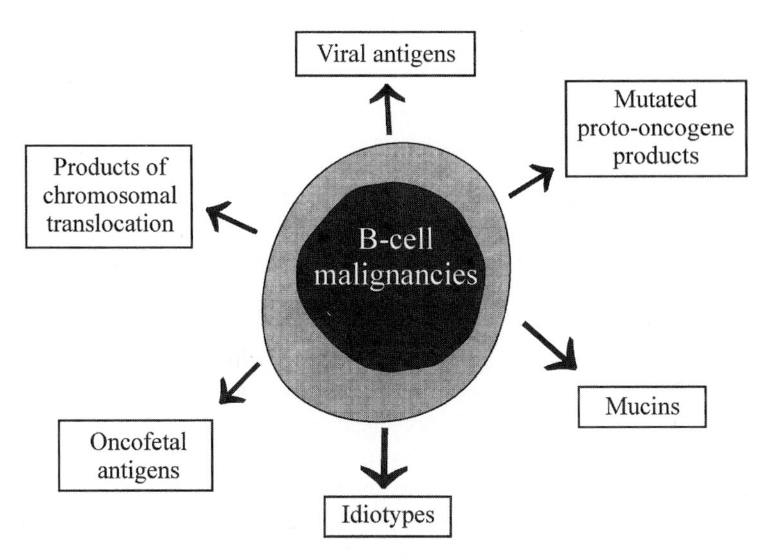

Fig. 1. Candidate tumor antigens expressed by B cell malignancies

In patients, B cell tumors have developed in spite of expressing several candidate tumor antigens. They also grow in sites which are exposed to the full power of the immune system. Evidently the antigens are not being presented in a manner likely to induce effective immunity, and the challenge is to activate a defeated immune response, in a situation where tumor antigens may still be present, and where the patient may be debilitated by disease or treatment. A potential problem is that tolerance may exist in the CD4+ T cell population, and many immunotherapeutic strategies are aimed at overcoming tolerance without inducing autoimmunity. In many cases, activation of immunity, for example by transfecting cytokine genes into tumor cells (Hock et al. 1993), is being attempted without knowledge of the tumor antigens involved. It is possible that this approach may reveal new antigens. We have taken the alternative approach of generating vaccines against defined antigens, using DNA delivery. While this limits the strategy to known sequences, it allows flexible design for optimal stimulation of an appropriate immune response.

8.2 DNA Vaccines – General

The first surprising observation was that injection of naked DNA containing the gene encoding β-galactosidase into muscle led to expression of functional enzyme (Wolff et al. 1990). This was followed by the discovery that encoded proteins derived from pathogens, delivered via this route, could induce antibody, CD4+ T cell responses, and cytotoxic T cells (CTLs) (Ulmer et al. 1993). Importantly, protective immunity could be elicited against infectious organisms, opening the way for generating simple sequence-dependent vaccines against a wide range of pathogens. It appears that the injected DNA does not integrate into the genome, but persists for long periods, unless the transfected muscle fibers are removed by CTL attack (Davis et al. 1997). Transcription of the encoded gene is usually driven by the CMV promoter, and activation of the immune system is assisted by the presence of immunostimulatory sequences within the plasmid backbone (Sato et al. 1996). The effect of these bacterial unmethylated CpG dinucleotide repeats is to induce production of cytokines, including IFNγ, IFNα, IFNβ, IL-12, and IL-18, and direct antibody responses down a TH1-dominated pathway (Klinman et al. 1996).

DNA vaccines are designed to deliver defined antigens to the immune system. However, to be widely applicable to cancer, they will need to activate immunity against a range of molecular categories. These include tumor antigens expressed at the cell surface, where antibody might be a key mediator of protection, and intracellular antigens expressed only as peptides associated with the MHC class I or class II molecules, where CD8+ or CD4+ T cell attack is the goal. In some cases, tumor antigens may be in a secreted form, and, although antibody may be predicted to be ineffective against these, it is not clear yet which T cells will be most useful. A powerful aspect of DNA vaccines is that they allow model systems to be developed where tumor antigens can be expressed in alternative molecular forms, so that rational designs can be tested.

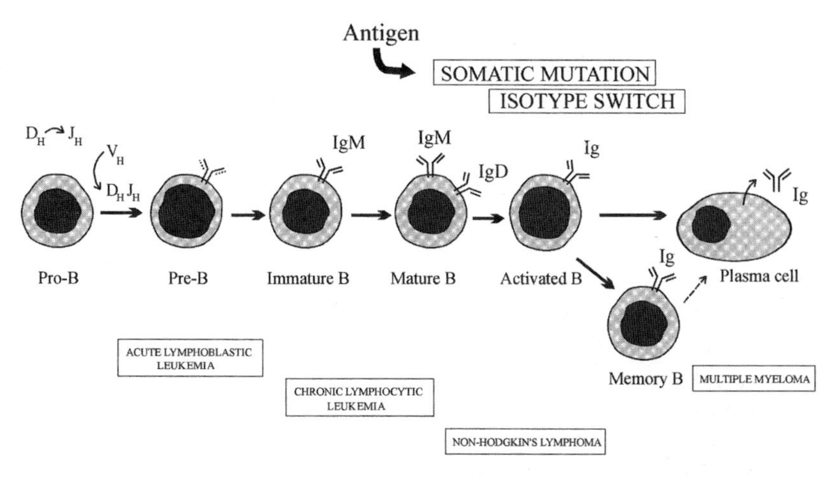

Fig. 2. Origin of B cell tumors in relation to normal B cell maturation. Immunoglobulin genes undergo recombination, somatic mutation, and isotype switching events during normal B cell differentiation. Immunogenetic analysis therefore provides an indicator of the point reached by the cell of origin prior to neoplastic transformation, and adds to tumor classification

8.3 Idiotypic Ig of B Cell Malignancies

Idiotypic determinants of B cell tumors arise from the Ig synthesized by the B cell of origin. Idiotypic Ig can be considered as a well-defined tumor antigen, but it is expressed in different molecular forms according to the maturation state of the B cell. This ranges from the μ chain of acute lymphoblastic leukemia, through the surface Ig expressed by B cell lymphomas, to the secreted Ig characteristic of multiple myeloma (Fig. 2). Generation of effective immunity against the whole range of B cell tumors therefore presents a challenge in vaccine design. Idiotypic determinants are encoded by the Ig variable region genes, V_H-D-J_H and V_L-J_L which have been assembled by genetic recombinatorial events, leading to sequences unique to each B cell. Further sequence diversity arises from somatic mutation, initiated following stimulation with antigen, usually in a germinal center environment. Lymphomas such as follicle center lymphoma, which develop in the germinal center and remain in that site, often continue to accumulate somatic mutations (reviewed in Stevenson et al. 1998). However, although the resulting changes could modulate the idiotypic determinants, they will be minor. Since the polyclonal immune response induced by vaccination is directed against the totality of determinants, these minor changes should have little effect.

8.4 Idiotypic DNA Vaccines Against Lymphoma

For B cell tumors, identification of V_H and V_L sequences used to encode idiotypic determinants is relatively simple by PCR/cloning and sequencing (Hawkins et al. 1994). We opted to assemble V-region genes in a single chain Fv (scFv) format, incorporating a 15 amino acid linker (Hawkins et al. 1994). We included a leader sequence for entry to the endoplasmic reticulum, and cloned the genes into a pcDNA3 vector for testing performance in inducing anti-Id antibodies. We soon found that scFv alone was poor in generating anti-Id immunity, and the reason for this has become clear. Figure 3 summarizes our current perception of the fate of encoded protein following injection into muscle or skin of a mouse. It is likely that long-lived muscle cells act as a depot of antigen (Davis et al. 1997), and that the fate of the antigen depends on the molecular form. Secreted antigen can be taken up and processed by APCs, whereas intracellular protein has to be released from muscle cells by a different route, possibly involving vesicle formation. This route can lead to induction of CTL responses and has been termed "cross priming" (Carbone and Bevan 1990).

With scFv alone as encoded antigen, we failed to induce a significant reproducible anti-Id response, and consequently were unable to activate protection against lymphoma (Stevenson et al. 1995). This contrasted with the finding that we could induce a response against scFv protein when injected with a powerful adjuvant, and suggested a need for additional stimulatory molecules. However, co-injection of DNA plasmids encoding GM-CSF, IL-2, or IFNγ did not raise the response to effective levels. Other groups were reporting similar difficulties using whole idiotypic Ig encoded in the plasmid, but had observed that immunogenicity could be improved by the presence of xenogeneic constant region sequence (Syrenglas et al. 1996). Our approach was to fuse a gene encoding a highly immunogenic protein, the fragment C (FrC) of

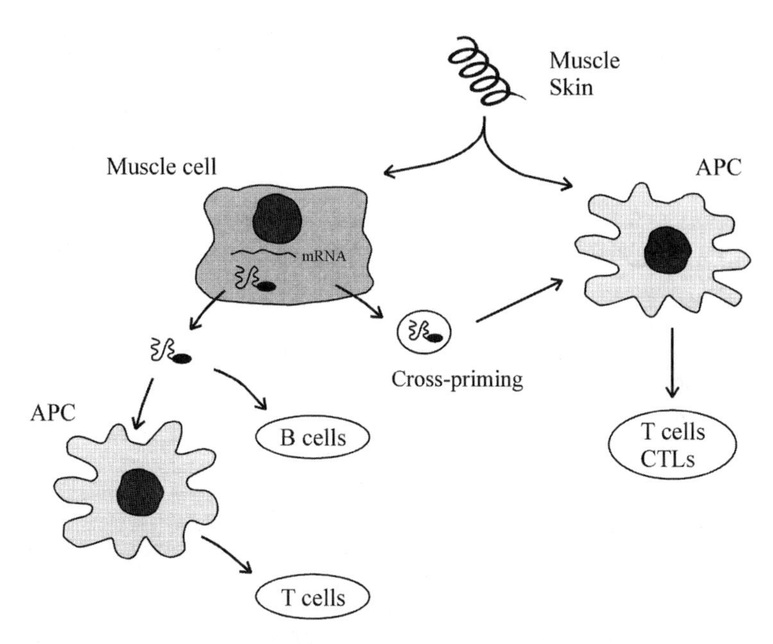

Fig. 3. Potential pathways of antigen presentation following DNA vaccination. DNA encoding single chain Fv (scFv) linked to fragment C (FrC) is injected into muscle or skin sites. Muscle cells act as an antigen depot for secretion of soluble scFv-FrC which is taken up by antigen-presenting cells (*APCs*) for presentation to CD4+ T cells. B cells may also take up scFv-FrC protein via specific surface Ig. Crosspriming from muscle cells is the route by which intracellular antigen reaches APCs, which can then activate cytotoxic T cells (*CTLs*). Direct transfection of APCs can occur, especially from skin sites

tetanus toxin, to the 3'-end of the scFv sequence. The fusion protein was expressed following transfection in vitro, and appears to be dimerized, with evidence for a minor component of multimers.

The results of injecting the fusion genes into mice were dramatic (Fig. 4), with scFv-FrC fusion genes constructed from four random patients' tumor Ig sequences inducing strong antibody responses, in each case against the patient's tumor Ig. In only one case (Fig. 4, WD), did scFv sequence alone generate a significant anti-Ig response. The other important factor was that the anti-Ig responses, although xenogeneic, were largely specific for the particular patient's tumor Ig, indi-

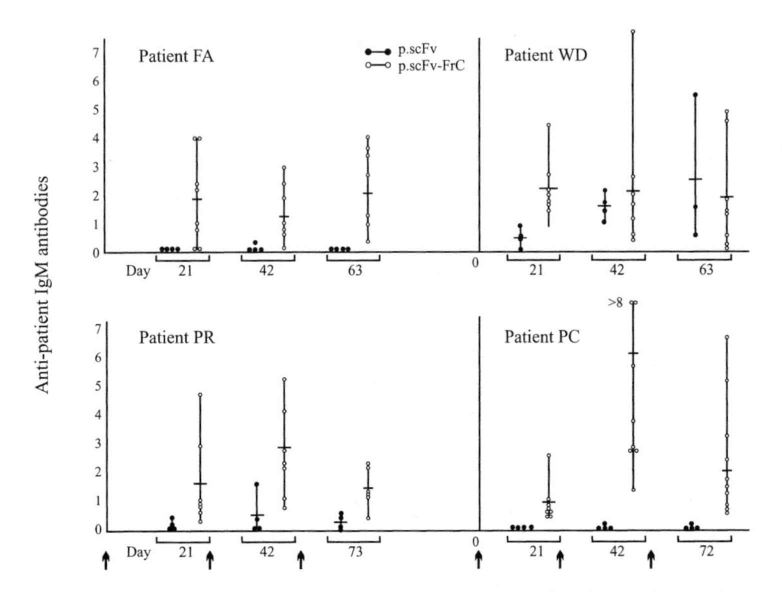

Fig. 4. Antibody responses against patients' tumor-derived IgM induced by DNA scFv vaccines. DNA scFv vaccines were assembled using V_H and V_L sequences from patients with B cell tumors. In each case, plasmids contained either scFv sequence alone (*p.scFv*) or fused to FrC sequence (*p.scFv-FrC*). Mice were injected in intramuscular sites with 50 µg DNA on days 0, 21, and 42, and bled on day 63. Antibodies against tumor IgM were measured by ELISA, and each *point* represents a single mouse

cating that the scFv-FrC fusion proteins were folding to display idiotypic determinants (Spellerberg et al. 1997).

It appears therefore that fusion of FrC sequence has rendered the scFv attractive to the immune system (Fig. 3). In order to understand the mechanism of this promotional effect, we tested the ability of separate plasmids encoding scFv alone and FrC alone to induce a response. The results were very clear: with separate vectors, a strong antibody response was induced against FrC but there was no promotion of the antibody response against the patient's scFv (King et al. 1998). Even when the two genes were within a single plasmid, no promotion was achieved. This indicates that there must be fusion between the two proteins, and the mechanism is likely to be a classic hapten-carrier

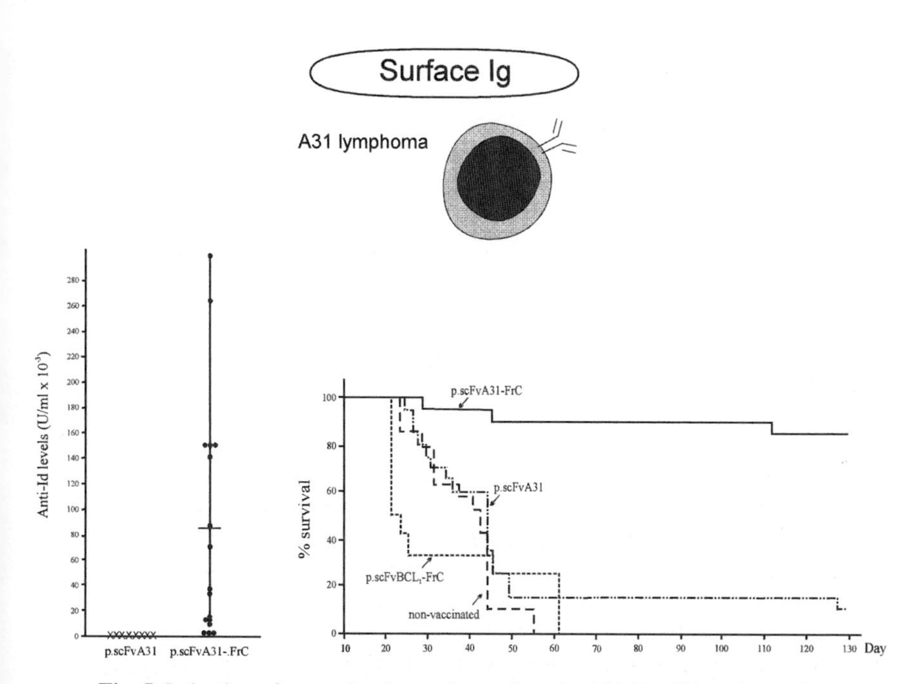

Fig. 5. Induction of protective immunity against the A31 B cell lymphoma by a DNAscFvA31-FrC vaccine. Mice were vaccinated with plasmids containing scFvA31-FrC (*p.scFvA31-FrC*) or scFv alone (*p.scFvA31*), on days 0, 21, and 42. A control fusion gene plasmid containing scFv derived from the BCL_1 lymphoma (*p.scFvBCL₁-FrC*) was also tested. Anti-idiotypic (Id) antibodies against A31 tumor IgM were measured and mice were challenged with tumor on day 63. Significant protective anti-Id responses were obtained only with the p.scFvA31-FrC fusion vaccine

effect, involving cognate T cell help to B cells secreting anti-Ig. This could be important in providing T cell help in a situation where the CD4+ T cells have been made tolerant by the long-term presence of tumor cells. FrC may also have other desirable properties as an immunogen, such as providing a danger signal to APCs which may promote uptake and processing of the fusion protein (Fig. 3).

The ability of the DNA scFv-FrC fusion gene vaccine to induce anti-Id responses in a syngeneic model was tested using the A31 mouse lymphoma (King et al. 1998). The results were similar to those found with human scFv, in that scFv alone was a poor inducer of anti-Id antibodies, but that fusion to FrC strongly promoted the response. Importantly, this led to protective immunity against lymphoma challenge (Fig. 5; King et al. 1998). Since we know from data from vaccination with idiotypic protein that anti-Id antibody is an effective mediator of protection (George et al. 1987; Kaminiski et al. 1987), it is likely that antibody is involved in protection after DNA vaccination.

8.5 DNA Fusion Gene Vaccines Against Alternative Tumor Antigens

The general applicability of the fusion gene approach to activating immune responses against cancer was tested using a completely different candidate tumor antigen, carcinoembryonic antigen (CEA). CEA is a 180-kDa glycoprotein expressed via a glycosyl phosphatidylinositol linkage on the surface of adenocarcinomas arising in a range of epithelial sites (Shively and Beatty 1985). It has an extended polypeptide structure highly decorated with oligosaccharides, and its main function is to act as an adhesion molecule. Using the first two domains of CEA in the construct, we tested the ability of CEA sequence alone or a CEA-FrC fusion gene, to induce anti-CEA responses. The results paralleled those found using scFv, in that CEA alone was a poor immunogen, and that this poor performance could be strongly promoted by fusion to FrC (Fig. 6). These findings open the possibility that the fusion gene design may be applicable to a wide variety of tumor antigens expressed at cell surfaces.

8.6 Idiotypic DNA Fusion Gene Vaccines Against Secreted Ig of Multiple Myeloma

Idiotypic Ig protein found in the serum of patients is readily available as a potential tumor antigen against myeloma. However, for myeloma, the challenge is to induce an immune response capable of killing neoplastic plasma cells which do not express surface Ig, but secrete large amounts. Idiotypic protein vaccines have been tested (Mellstedt and Osterborg 1999), but one requirement will be to deliver the antigen via a route which can induce T_H1 or CTL responses. To test the ability of idiotypic

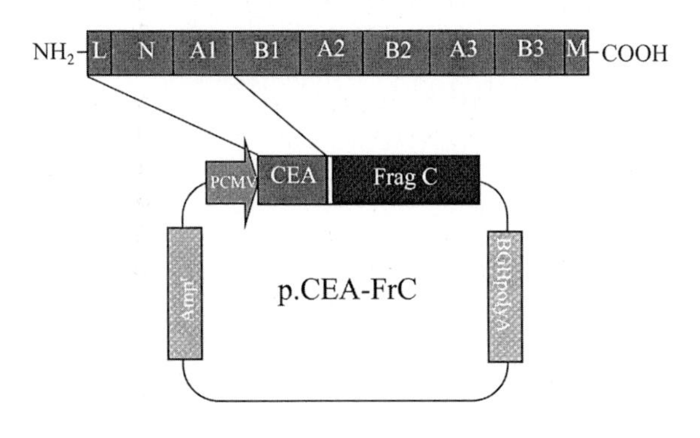

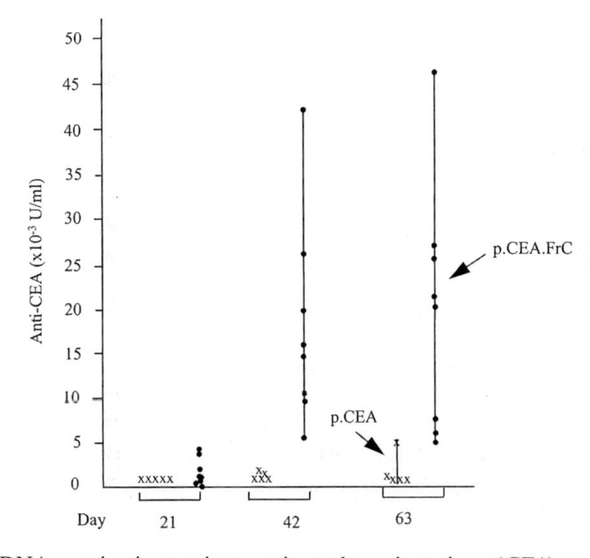

Fig. 6. DNA vaccination against carcinoembryonic antigen (*CEA*): promotion of anti-CEA antibody response by fusion of CEA sequence to FrC. The gene encoding the two N-terminal domains of CEA was fused to FrC sequence and used in a DNA vaccine format. Mice were injected with plasmids containing either CEA sequence alone (*p.CEA*) or fused to FrC sequence (*p.CEA-FrC*), on days 0, 21, and 42. Antibodies against CEA were measured by ELISA in sera taken on days 21, 42, and 63. Amplification of the antibody response was seen with the fusion construct

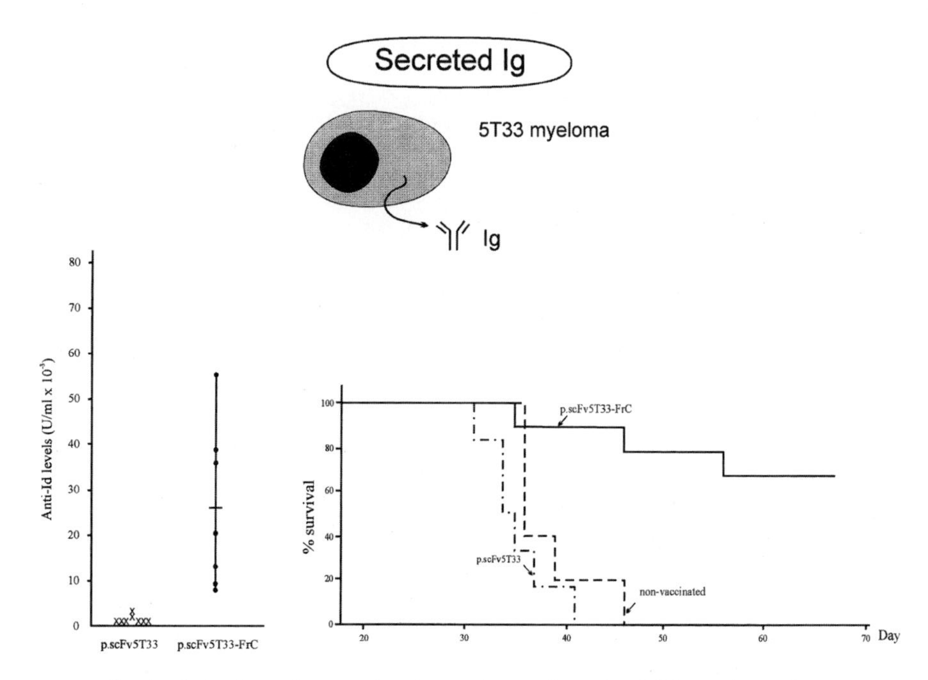

Fig. 7. Induction of protective immunity against the 5T33 myeloma by a DNAscFv5T33-FrC vaccine. Mice were vaccinated with plasmids containing scFv5T33-FrC (*p.scFv5T33-FrC*) or scFv alone (*p.scFv5T33*), on days 0, 21, and 42. Anti-Id antibodies against 5T33 myeloma IgG were measured and mice were challenged with tumor on day 63. Significant protective anti-Id responses were obtained only with the p.scFv5T33-FrC fusion vaccine

DNA vaccines to suppress myeloma, we used the 5T33 myeloma model developed by Dr. J. Radl (Croese et al. 1991). We identified and assembled the VH and Vk genes used to encode the idiotypic determinants, and tested the ability of constructs containing scFv sequence alone or the scFv-FrC fusion gene, to induce anti-Id antibodies in mice. As expected, no response was seen with scFv, but a promotional effect was found using the fusion gene (Fig. 7). Protection was also evident against challenge with myeloma, indicating that a surface Ig-negative tumor could be attacked by cells activated by the DNA vaccine. We have established that anti-Id antibody plays no role in protection, since vaccination with idiotypic protein with adjuvant raised high levels of anti-Id antibodies, but failed to generate protective immunity (King et al. 1998). No CTL activity against the myeloma cells could be detected, and we are currently investigating the involvement of CD4+ T cells as likely mediators of protection.

8.7 Effect of Pre-existing Anti-FrC Antibody

Prior to using the scFv-FrC fusion gene in patients, we were concerned to know if pre-existing anti-FrC antibodies would suppress the subsequent response to scFv. This concern arose from the observation that epitope suppression can occur against peptides linked to tetanus toxoid (TT) as a carrier when anti-TT antibodies are already present (Panina-Bordignon et al. 1989). To test this, we prevaccinated mice with TT and investigated the effect of this on induction of immunity against scFv-FrC. We found that the levels of anti-FrC antibodies characteristic of patients had no significant effect on the subsequent response to scFv, but that if levels were increased beyond this by vaccinating immediately prior to DNA injection, suppression of anti-Id responses could occur. This needs to be taken into account for a clinical trial, but should not present a major problem.

8.8 Clinical Trial of DNA Vaccines

A phase I clinical trial of idiotypic DNA scFv vaccines in patients with end-stage low-grade follicle cell lymphoma (FCL) has been carried out in seven patients (Hawkins et al. 1997). Since the patients were profoundly immunosuppressed, this was largely a toxicity study, and no serious side effects were observed following intramuscular injection of escalating doses of DNA. A second trial of DNA scFv-FrC fusion gene vaccines is beginning with doses again increasing from 300 to 2400 μg. Importantly, patients will be with FCL in first remission, and treatment with purine analogues will be an exclusion criterion. The major end-point will be detection of antibody against both FrC and against the patient's idiotypic protein. If responses are obtained, we shall apply to extend the trial to patients with myeloma.

8.9 DNA Vaccines Against Intracellular Tumor Antigens

So far, we have developed DNA vaccine designs capable of inducing protective immunity against surface or secreted antigens. However, the bulk of candidate tumor antigens consists of intracellular antigens which are only visible to the immune system as peptides in the groove of the MHC class I molecules (Rosenberg et al. 1996). Attack on these antigens is likely to require direct action by CD8+ T cells, and many antigens have been defined, particularly in patients with melanoma, by isolation of CTLs (van den Eynde et al. 1995). However, there are possible routes for attack by CD4+ T cells, either indirectly via cytokine release or directly by an as yet undefined pathway (Wise et al. 1999).

In order to design DNA vaccines to induce T cell-mediated pathways of attack, we have generated a tumor model consisting of EL-4 tumor cells into which we have transfected the gene encoding a non-secreted form of FrC. This tumor can be passaged in mice with no spontaneous induction of protective immunity. However, vaccination with DNA encoding FrC can induce specific antibody, CD4+ and CD8+ T cell responses. In addition, there is protection against tumor challenge. This model is allowing insight into the nature of the desired protective immunity against tumor and the means to achieve it.

8.10 Concluding Remarks

Molecular genetic technology is facilitating many new approaches to the understanding and rational treatment of cancer. However, it took approximately 200 years from Jenner's early vaccination against smallpox to reach total eradication of the disease, and we shall need time to develop effective vaccines against cancer. DNA vaccines offer opportunities to activate all arms of the immune response, and the ease of construction and manipulation means that new designs can be tested quickly. These vaccines are already in clinical trials for prevention of certain infectious diseases (Wang et al. 1998), and we shall learn a great deal from these.

For cancer, immunity usually has to be induced in patients with disease, and success is likely to depend on first achieving clinical remission. Ideally, this will allow recovery of immune capacity without resur-

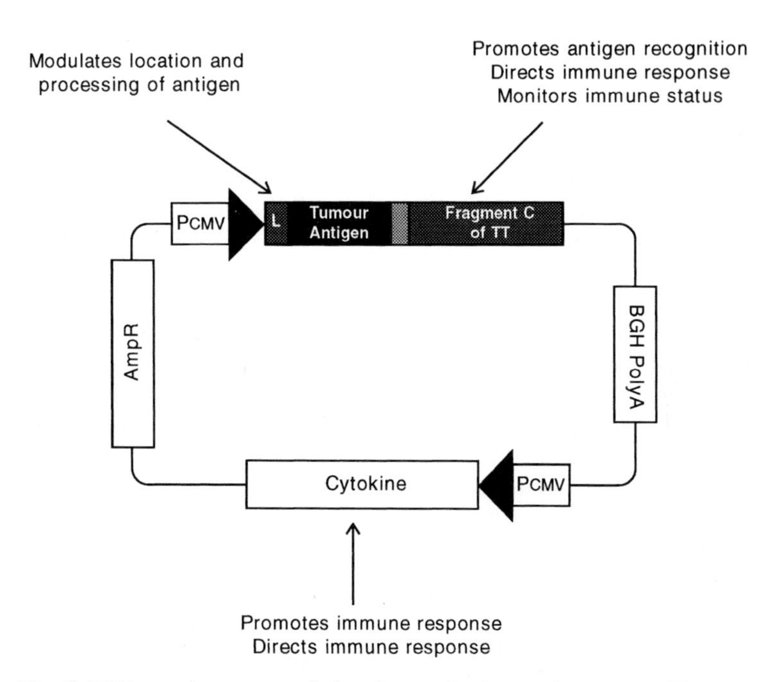

Fig. 8. DNA vaccine cassette design for vaccination against cancer. The transcriptional unit consists of a gene encoding a tumor antigen (scFv or CEA) fused to the FrC sequence. Expression is driven by the CMV promoter and there is a leader sequence to allow entry of protein into the endoplasmic reticulum. There is an opportunity to incorporate genes encoding cytokines to amplify or direct the immune response, if required

gent tumor. The potential problem of tolerance may be overcome by transplantation, and, in the setting of allogeneic transplantation, there is the tempting possibility of vaccinating the donor. However, additional CD4+ T cell help may be required and this can be provided by adjuvant sequences in the vaccine. Activation of dendritic cells by stimuli such as CD40L or anti-CD40 antibodies may also bypass the requirement for T cell help (Grewal and Flavell 1998), and these can be delivered via DNA (Mendoza et al. 1997).

The principle of activating immunity against cell surface or secreted antigens by DNA vaccines encoding antigen linked to FrC should apply to a range of tumor antigens, and many are being defined. The aim is to produce DNA vaccine cassettes which can be patient-specific or tumor-specific (Fig. 8). These are simple in concept but provide flexibility in design. If necessary, genes encoding cytokines could be incorporated into the cassette. For intracellular antigens such as mutated proto-oncogenes, a similar principle may be applicable but detailed strategy will differ. There are exciting prospects for harnessing the information and technology provided by genetics and we are poised to translate these into new treatments for cancer.

Acknowledgements. This work was supported by a Leukaemia Research Fund Specialist Programme Grant, Tenovus, Cardiff, and the Cancer Research Campaign, UK. We also thank the Kathy Giusti Multiple Myeloma Research Foundation for support.

References

Carbone FR, Bevan MJ (1990) Class-1 restricted processing of exogenous cell-associated antigen in vivo. J Exp Med 171:377–387

Croese JW, Vissinga CS, Boersma WJA, Radl J (1991) Immune regulation of mouse 5T2 multiple myeloma. Immune response to 5T2 MM idiotype. Neoplasma 38:457–466

Davis HL, Brazolot Millan CL, Watkins SC (1997) Immune-mediated destruction of transfected muscle fibers after direct gene transfer with antigen-expressing plasmid DNA. Gene Ther 4:181–188

Eynde B van den, Peeters O, Backer O de, Gangler B, Lucas S, Boon T (1995) A new family of genes coding for an antigen recognized by autologous cytolytic T lymphocytes on a human melanoma. J Exp Med 182:689–698

George AJT, Stevenson FK (1989) Prospects for the treatment of B cell tumors using idiotypic vaccination. Int Rev Immunol 4:271–310

George AJT, Tutt AL, Stevenson FK (1987) Anti-idiotypic mechanisms involved in suppression of a mouse B cell lymphoma, BCL. J Immunol 138:628–634

Grewal IS, Flavell RA (1998) CD40 and CD154 in all mediated immunity. Ann Rev Immunol 16:111–135

Hawkins RE, Zhu D, Ovecka M, Winter G, Hamblin TJ, Long A, Stevenson FK (1994) Idiotypic vaccination against human B-cell lymphoma. Rescue of variable region gene sequences from biopsy material for assembly as single-chain Fv personal vaccines. Blood 83:3279–3288

Hawkins RE, Stevenson FK, Russell SJ, Hamblin TJ (1997) A pilot study of idiotypic vaccination for follicular B-cell lymphoma using a genetic approach. Hum Gene Ther 8:1287–1299

Hock H, Dorsch M, Kunzendorf U, Überla K, Qin Z, Diamantstein T, Blankenstein T (1993) Vaccinations with tumor cells genetically engineered to produce different cytokines: effectivity not superior to a chemical adjuvant. Cancer Res 53:714–716

Kaminiski HS, Kitamura K, Maloney DG, Levy R (1987) Idiotypic vaccination against a murine B cell lymphoma. Inhibition of tumor immunity by free idiotypic protein. J Immunol 138:1289–1296

King CA, Spellerberg MB, Zhu D, Rice J, Sahota S, Thompsett AR, Hamblin TJ, Radl J, Stevenson FK (1998) DNA vaccines with single chain Fv fused to fragment C of tetanus toxin induce protective immunity against lymphoma and myeloma. Nat Med 4:1281–1286

Klinman DM, Yi A, Beaucage SL, Conover J, Krieg AM (1996) CpG motifs expressed by bacterial DNA rapidly induce lymphocytes to secrete IL-6, IL-12 and IFN-γ. Proc Natl Acad Sci USA 93:2879–2883

Mellstedt H, Osterborg A (1999) Active idiotypic vaccination in multiple myeloma. GM-CSF may be an important adjuvant cytokine. Pathol Biol 47:211–215

Mendoza RB, Cantwell MJ, Kipps TJ (1997) Immunostimulatory effects of a plasmid expressing CD40 ligand (CD154) on gene immunization. J Immunol 159:5777–5781

Nigro L, Baker SJ, Preisinger AC, Jessup JM, Hostetter R, Cleary K, Bigner SH, Davidson N, Baylin S, Devilee P (1989) Mutations in the p53 gene occur in diverse human tumor types. Nature 342:705–707

Panina-Bordignon P, Tau A, Termijtelen A, et al (1989) Universally immunogenic T cell epitopes: promiscuous binding to human MHC class II and promiscuous recognition by T cells. Eur J Immunol 19:2237–2242

Rabbitts TH (1994) Chromosomal translocations in human cancer. Nature 372:143–149

Racila E, Hseuh R, Marches R, Tucker PF, Krammer PH, Scheuermann RH, Uhr JW (1996) Tumor dormancy and cell signalling. IV. Anti-μ induced apoptosis in human B lymphoma cells is not caused by an APO-1-APO-1 ligand interaction. Proc Natl Acad Sci USA 93:2165–2168

Rosenberg SA, Kawakami Y, Robbins PF, Wang R (1996) Identification of the genes encoding cancer antigens: implications for cancer immunotherapy. Adv Cancer Res 70:145–177

Sato Y, Roman M, Tighe H, Lee D, Corr M, Nguyen MD, Sliverman GJ, Lotz M, Carson DA, Raz E (1996) Non-coding bacterial DNA sequences necessary for effective intradermal gene immunization. Science 273:352–354

Shively J, Beatty J (1985) CEA related antigens: molecular biology and clinical significance. Crit Rev Oncol Hematol 2:355–399

Spellerberg MB, Zhu D, Thompsett A, King CA, Hamblin TJ, Stevenson FK (1997) DNA vaccines against lymphoma. Promotion of anti-idiotypic antibody responses induced by single chain Fv genes by fusion to tetanus toxin fragment C. J Immunol 159:1885–1892

Stevenson FK, Zhu D, King CA, Ashworth LJ, Kumar S, Hawkins RE (1995) Idiotypic vaccines against B-cell lymphoma. Immunol Rev 145:211–228

Stevenson FK, Sahota S, Zhu D, Ottensmeier C, Chapman C, Oscier D, Hamblin T (1998) Insight into the origin and clonal history of B-cell tumors as revealed by analysis of immunoglobulin variable region genes. Immunol Rev 162:247–259

Syrenglas AD, Chen TT, Levy R (1996) DNA immunization induces protective immunity against B cell lymphoma. Nat Med 2:1038–1041

Takahashi T, Makiguchi Y, Hinoda Y, Kakiuchi H, Nakagawa N, Imai K, Yachi A (1994) Expression of MUC1 on myeloma cells and induction of HLA-unrestricted CTL against MUC1 from a multiple myeloma patient. J Immunol 153:2102–2109

Ulmer JB, Donnelly JJ, Parker SE, Rhodes GH, Felgner PL, Dwarki VJ, Gromkowski SH, Deck RR, DeWitt CM, Friedman A (1993) Heterologous protection against influenza by injection of DNA encoding a viral protein. Science 259:1745–1749

Van Baren N, Brasseur F, Godelaine D, Hames G, Ferrant A, Lehmann F, Andre M, Ravoet C, Doyen C, Spagnoli GC, Bakkus M, Thielemans K, Boon T (1999) Genes encoding tumor-specific antigens are expressed in human myeloma cells. Blood 94:1156–1164

Wang R, Doolan DL, Le TP, et al (1998) Induction of antigen-specific cytoxic T lymphocytes in humans by a malaria DNA vaccine. Science 282:476–480

Wise M, Zelenika D, Bemelman F, Latinne D, Bazin H, Cobbold S, Waldmann H (1999) CD4 T cells can reject major histocompatibility complex class I-incompatible skin grafts. Eur J Immunol 29:156–167

Wolff JA, Malone RW, Williams P, Chang W, Acsadi G, Jani A, Felgner PL (1990) Direct gene transfer into mouse muscle in vivo. Science 247:1465–1468

9 Skin Cancer – Prospects for Novel Therapeutic Approaches

G. Stingl

9.1 Introduction

The skin is a histogenetically diverse organ. This is also reflected in the type of malignancies originating from it which range from epithelial (basal cell carcinoma, squamous cell carcinoma), neuroectodermal (for example, melanoma), myeloid (for example, cutaneous T cell lymphoma, mastocytoma) to mesenchymal (for example, Kaposi's sarcoma) neoplasms.

In the past decade, we have made enormous progress in our understanding of events finally resulting in tumorigenesis and, conversely, of host-defense mechanisms counteracting them. Important examples include the demonstration of mutations in the human patched gene in patients with hereditary and sporadic basal cell carcinomas (Johnson et al. 1996; Unden et al.1997), the role of a p53 polymorphism in the development of human papilloma virus-associated cancer (Storey et al. 1998), the identification and characterization of several melanoma cell-associated and -specific antigens (Boon et al 1994), the identification of cytokines promoting the growth of Sézary lymphoma cells (Dalloul et al. 1992) and, last but not least, the discovery of the human herpesvirus 8 as the causative agent of Kaposi's sarcoma (Moore and Chang 1995). It is quite obvious that these and other achievements will lead to new strategies in the prevention and treatment of skin cancer and cancer in general. It is imperative that dermatologists know about these developments if they wish to actively participate in the care of skin cancer patients in the new millennium.

The coverage of all the different therapeutic approaches in cutaneous oncology would clearly exceed the scope of this article. Instead, I will focus my attention on one particular topic in this area, i.e., the design and production of therapeutic cancer vaccines with a special emphasis on melanoma, the leading cause of skin cancer deaths and, at the same time, the malignant neoplasm with the highest rate of spontaneous regression.

9.2 Protective and Nonprotective Cancer Immunity

Melanoma is the prototype of a cancer capable of inducing a productive and sometimes even protective anti-tumor immune response. This is best evidenced by the occasional occurrence of spontaneous melanoma regression (Fig. 1) which is often accompanied by the development of halo nevi, uveitis, and/or vitiligo. Although an important role of a tumor cell-specific antibody response in this process cannot be formally excluded, it appears that $CD4^+$ as well as $CD8^+$ T cells are the main carriers of a clinically relevant anti-melanoma immune response (Topalian et al. 1989; Mackensen et al. 1994). Some of the antigens recognized by the patients' tumor-infiltrating lymphocytes (TIL) have recently been identified and molecularly characterized (Boon et al. 1994). These include tumor-specific neoantigens (for example, MAGE family), tissue-specific differentiation antigens (for example, MART-1/melan-A, gp100/Pmel17, tyrosinase), and products of mutated genes (for example, CDK4, β-catenin, Ras).

All this should not detract from the fact that, in most instances, this endogenous anti-cancer immunity is not effective, and it may even happen that tumor cell-specific T cell tolerance develops. What determines the outcome of an immune response is the context in which the

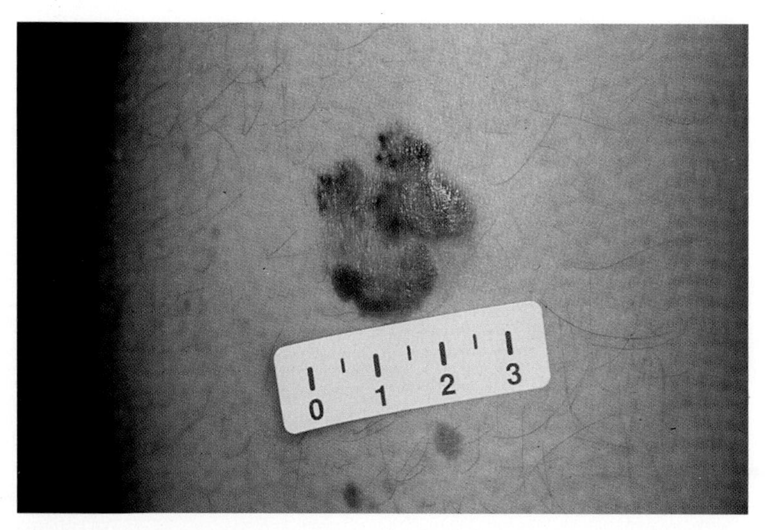

Fig. 1. Superficial spreading melanoma with clinical signs of regression

antigen is presented to the lymphocyte. If the T cell encounters an antigen-presenting cell (APC) that expresses the antigenic MHC/peptide complex as well as a panel of costimulatory molecules, the typical result is activation (Fig. 2). In the absence of the appropriate costimulatory signals, however, engagement of the T cell receptor leads to ignorance, anergy, or apoptosis of the antigen-specific T cell (Fig. 2; Mueller et al. 1989; van Parijs and Abbas 1998).

Upregulation of APC-bound costimulatory molecules is induced by proinflammatory cytokines and typically occurs after the delivery of a danger signal (for example, infection; Matzinger 1994). Since the various steps of tumorigenesis, including the display of neoantigens, usually occur in the absence of inflammatory events or tissue-destructive processes, it is not unreasonable to assume that tumor cell-associated antigens are either ignored by the immune system (Wick et al. 1997) or, for the worse, induce specific tolerance in T cells (Staveley-O'Carroll et al. 1998; Lee et al. 1999). Thus, in order to be effective, cancer vaccines must be capable of either evoking a productive immune response in tumor antigen-reactive T cells previously ignored or, what is even more challenging, of breaking tolerance.

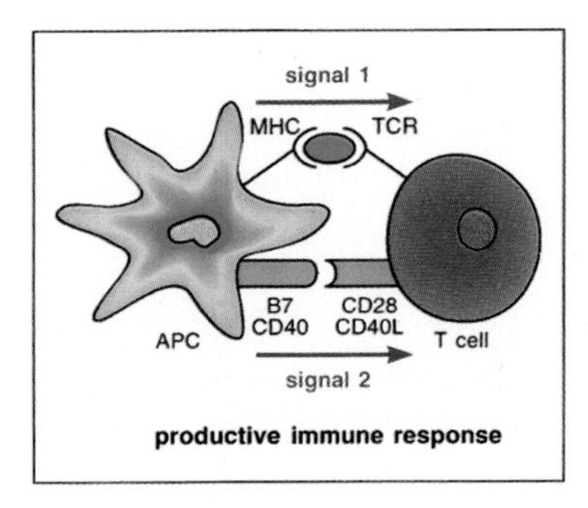

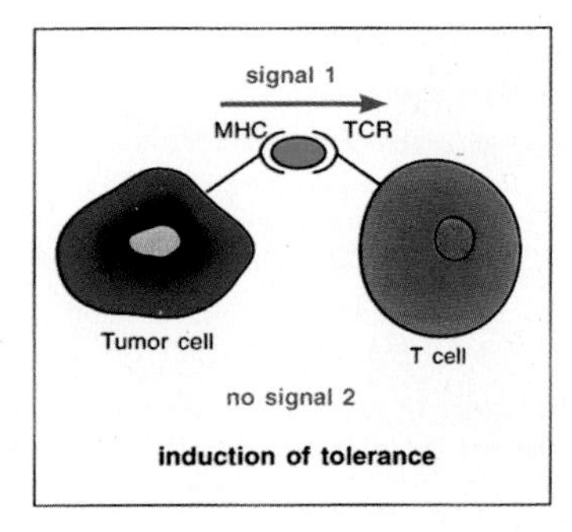

Fig. 2. Both professional (*left*) and non-professional (*right*) antigen-presenting cells (APCs) express the MHC-bound antigenic peptide. The presence (*left*) or absence (*right*) of costimulatory molecules determines the final outcome of the immune response

This can only be accomplished when the appropriate tumor-associated antigens are targeted to and, as a consequence, processed and presented by potent professional APCs such as dendritic cells (DCs; Steinman 1991; Cella and Lanzavecchia 1997). These cells reside within non-lymphoid and lymphoid tissues and, upon an appropriate encounter, can capture and process soluble protein antigens (Crowley et al. 1990). Only after receipt of danger signals provided by cytokines or bacterial products do they switch their machinery to an immunostimulatory mode. This is evidenced by the surface display of large numbers of long-lived peptide-loaded MHC antigens and costimulatory molecules as well as by their capacity to migrate.

Tumor antigens can be targeted to APCs in several ways. Apoptotic cancer cells and fragments thereof as well as large protein antigens can be endocytosed by APCs and are then channeled not only in their MHC class II, but also in their MHC class I processing pathway (Huang et al. 1994; Albert et al. 1998). This phenomenon, termed crosspriming, provides the rational basis for the use of allogeneic cancer cell vaccines (see below). An alternative possibility for expressing tumor-associated antigens in APCs is to transfect these cells with genes encoding such molecules. This mechanism is probably operative in certain recombinant viral and bacterial vaccines as well as in naked DNA vaccines. Finally, the possibility exists to directly load APC-bound empty MHC molecules with tumor antigen-derived peptides (Hu et al. 1996; Buschle et al. 1997).

9.3 Immunotherapy of Melanoma

One strategy to immunologically combat cancer is to administer tumor cell-specific effector molecules and/or cells to the patients. In cancers other than melanoma, we are currently experiencing a renaissance of clinical trials using (humanized) cancer cell-specific monoclonal antibodies (Riethmüller et al. 1994; Katsumata et al. 1995; Coiffier et al. 1998).

In melanoma, Rosenberg and coworkers (1988) pioneered studies with TIL isolated and grown from metastatic melanoma nodules. Adoptive transfer of these cells, together with high-dose interleukin 2 (IL-2), resulted not infrequently in, at least, partial tumor remissions but, because of the frequency and severity of adverse reactions, this treatment did not gain wide acceptance among clinical oncologists.

Therefore, considerable efforts have been made to develop cancer vaccines capable of inducing and/or enhancing anti-melanoma cell immune responses that, under ideal circumstances, should lead to cancer cell destruction. Currently, we must admit that, despite impressive and spectacular vaccine-induced cancer remissions in experimental animals, the value of cancer vaccines for the treatment of human neoplasms has yet to be demonstrated.

9.3.1 Cancer Cell-Based Vaccines (Fig. 3)

Although major progress has been made in the definition and characterization of immunologically relevant cancer antigens, our knowledge about the clinically most relevant tumor rejection antigens is still very limited. For this reason, most cancer vaccine approaches still use tumor cells themselves or fragments thereof as a source of antigen. In order to increase their immunogenicity, cancer cells or cancer cell lysates are either administered together with adjuvants (Elias et al. 1997), fused with professional APCs such as DCs (Gong et al. 1997), or genetically modified before being injected into the patient. In the case of melanoma,

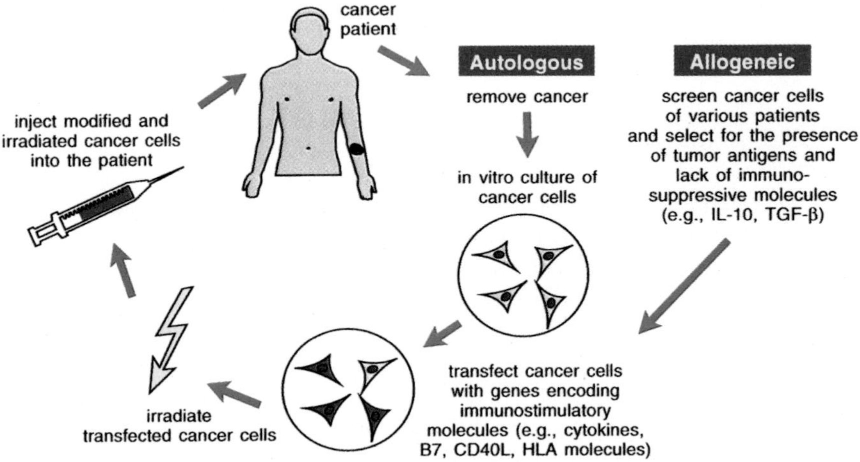

Fig. 3. Cancer cells can be genetically modified to express immunostimulatory molecules. When injected into the patient, they elicit a self-destructive inflammatory response. Tumor cell fragments thus generated can be processed by host APCs and be presented to T cells in an immunologically relevant fashion

evidence exists that treatment with lysates of vaccinia-infected melanoma cells can improve the survival of patients with stage II (old nomenclature) disease (Hersey et al. 1990). Several years later, it was shown that transfection of murine melanoma cells with genes encoding cytokines (Gänsbacher et al. 1990; Zatloukal et al. 1995) or costimulatory molecules (Townsend and Allison 1993) reduces their tumorigenicity and, conversely, increases their immunogenicity. This vaccine-induced cancer immunity is biologically relevant, i.e., is capable of eliminating wild-type cancer cells in a prophylactic and, to a somewhat lesser extent, therapeutic setting (Zatloukal et al. 1995).

It appears that most cytokine-based cancer cell vaccines do not stimulate T cells directly but rather by inducing a cytodestructive inflammatory response and, subsequently, the presentation of tumor cell fragments generated by host APCs (Huang et al. 1994). In the case of IL-2-based melanoma cell vaccines, natural killer (NK) cells apparently play an important role in the lysis of the genetically modified cancer cells (Schneeberger et al. 1999).

Several clinical phase I trials with cytokine-expressing autologous cancer cells have been conducted in patients with far advanced melanoma (Abdel-Wahab et al. 1997; Möller et al. 1998; Schreiber et al. 1999) and other types of cancer (Simons et al. 1997). Autologous cancer cell vaccines, particularly when consisting of tumor cells from different sites, express the entire spectrum of cancer cell-associated antigenic specificities of a given patient and may thus seem to be ideally suited for inducing clinically effective anti-cancer immunity. We should, however, not forget that autologous vaccines, apart from being expensive and labor-intensive in their preparation, can hardly be produced in a standardized fashion and can therefore not be reliably tested for immunological or, even, clinical efficacy. One possibility to resolve this problem is to use irradiated unmodified autologous cancer cells together with gene-modified bystander cells (Veelken et al. 1997) and/or biopolymer microspheres (Golumbek et al. 1993) producing and/or releasing the respective cytokine in a controlled fashion. The currently most widely used approach involves standardized gene-transfected cancer cell lines selected and generated from different patients on the basis of their growth characteristics, antigenic repertoire, and cytokine-secreting properties. The rationale for this form of treatment came from the demonstration that T cell crosspriming, i.e., the presentation of antigenic specificities of an allogeneic cancer cell by host APCs, can occur in vivo (Huang et al. 1994) and that MHC compatibility between patient and cancer cell is therefore not necessarily needed for this vaccine to display its desired effect.

At this moment, we know that both autologous (Abdel-Wahab et al. 1997; Schreiber et al. 1999) and allogeneic (Belli et al. 1997) cytokine-based melanoma cell vaccines are generally well tolerated. Data so far available do not allow us to make valid conclusions about their clinical or, even, immunological efficacy. This is partly due to difficulties in adequately measuring and monitoring immune responses possibly induced by an antigenically highly diverse vaccine.

9.3.2 Antigen-Specific Cancer Vaccines

Our rapidly increasing knowledge about structure as well as mode of presentation and recognition of melanoma-associated antigens led to the development of candidate antigen-specific vaccines for clinical testing.

9.3.2.1 Peptide Vaccines

Because most tumor-associated antigens were originally detected by their serving as targets of $CD8^+$ MHC class I-restricted cytotoxic T cells (Boon et al. 1994), most peptide-based cancer vaccines tested so far have used MHC class I-restricted peptide antigens. Upon injection, such peptides will bind to empty MHC class I molecules on both professional and non-professional APCs. The former event is desirable and should lead to a productive and protective anti-tumor immune response (Fig. 2; Hu et al. 1996; Buschle et al. 1997). The latter situation is potentially dangerous as it could result in cancer-specific tolerance (Fig. 2; Toes et al. 1996).

Peptide vaccines for melanoma are now being tested clinically (Hu et al. 1996; Jaeger et al. 1996), but data that unequivocally prove their clinical and/or immunological efficacy are still missing. In fact, the reliable assessment of a clinically relevant anti-melanoma immune response is a difficult task as we have learned that peptide vaccination may be followed by impressive clinical remissions but not by the development of peptide-specific CTL responses and vice versa (Marchand et al. 1999).

A possible strategy to increase the success rate of peptide vaccination could be to use anchor-modified peptides, i.e., peptide analogues which bind with much higher affinity to the respective HLA antigen and are much better recognized by TIL than the HLA-bound natural peptide. A study by Rosenberg et al. (1998a) indicates that a sizable percentage of HLA-A2-positive patients with far advanced melanoma took benefit from a vaccination with an immunodominant peptide from the gp100 melanoma-associated antigen, administered in incomplete Freund's adjuvant and together with IL-2.

An alternative approach for increasing the immunogenicity of a given tumor cell-associated antigenic peptide is to present it in the context of professional APCs (Fig. 4). The feasibility of this modality can be deduced from a recent study by Nestle et al. (1998) who repeat-

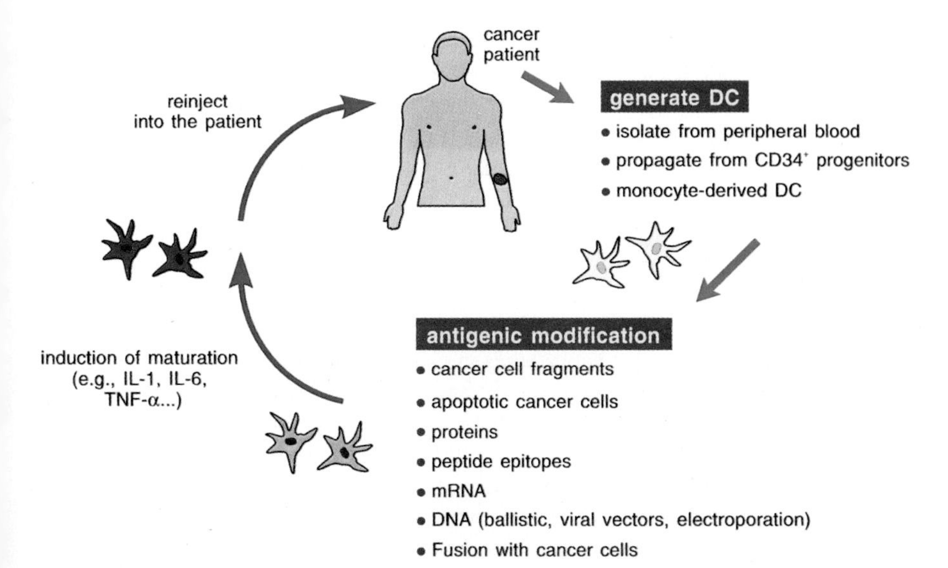

Fig. 4. Dendritic cells (DCs), the most potent APCs, can be either isolated from the peripheral blood or be generated in large numbers from either $CD34^+$ hematopoietic or monocytic precursors with appropriate cytokines. Expression of antigenic tumor peptides by these cells can be accomplished by direct peptide loading, by channeling tumor cell fragments, tumor-associated proteins, or apoptotic cancer cells into the MHC class I and class II presentation pathway, by gene transfer, or by fusion with cancer cells. Upon further maturation with proinflammatory cytokines (IL-1, TNF-α, IL-6), tumor antigen-expressing DCs can be used for vaccination purposes

edly injected monocyte-derived DCs (Sallusto and Lanzavecchia 1994) that had been pulsed with either tumor lysates or a peptide cocktail as well as with the helper antigen keyhole limpet hemocyanin (KLH) into clinically uninvolved lymph nodes of patients with advanced melanoma. Vaccination was well tolerated by all 16 recipients. They all developed delayed-type hypersensitivity (DTH) reactions toward KLH and 11/16 to peptide-pulsed DC. The antigenic specificity and potency of this DC vaccination protocol is further evidenced by the recruitment and expansion of peptide-specific CTLs from the DTH challenge site and by the demonstration of objective clinical remissions in 5 of the 16 patients.

Studies by Zitvogel et al. (1998) indicate that whole DCs may not even be required for DC-based cancer peptide vaccination to succeed. In fact, DCs can secrete antigen-presenting vesicles, termed exosomes, which contain MHC class I and class II molecules as well as costimulatory molecules. The observation that tumor peptide-pulsed DC-derived exosomes can prime specific CTLs in vivo and eradicate or suppress the growth of established tumors provides the important basis for testing this interesting form of immunotherapy in humans.

9.3.2.2 Recombinant and Nucleic Acid Vaccines

In the past few years, many studies in experimental animals have documented the potency of genetically engineered (recombinant) bacteria (Pan et al. 1995) and viruses (Wang et al. 1995), as well as various nucleic acid constructs, in inducing anti-cancer immune responses. Recombinant bacteria become increasingly attractive either because they possess enteric routes of infection (for example, *Salmonella*, *Listeria monocytogenes*) and thus allow for oral vaccine delivery, because they promote the survival and maturation of DCs (Rescigno et al. 1998a) and/or because they deliver exogenous antigens for efficient presentation not only in the MHC class II, but also in the MHC class I processing pathway (Rescigno et al. 1998b).

The generation of recombinant viral vaccines involves the insertion of genes encoding tumor antigens or immunodominant epitopes thereof into the viral genome. Their immunogenicity can be enhanced and/or modulated by the incorporation of genetic sequences promoting either MHC class I (McCabe et al. 1995) or MHC class II (Wu et al. 1995) presentation, of genes encoding immunostimulatory molecules (Chamberlain et al. 1996), or by selecting a promoter favoring gene expression in DCs (Bronte et al. 1997).

Clinical trials with recombinant viral cancer vaccines have begun recently (Borysiewicz et al. 1996; Rosenberg et al. 1998b). From the viewpoint of safety, the selection of viruses incapable of replicating in mammalians (for example, avian poxviruses) is desirable. Alternatively, recombinant viruses can be made safe by the removal of genes that are critical to replication and virulence.

A major impediment in the development of recombinant viruses for clinical use is that of pre-existing immunity of the recipients against the viruses most frequently used as vector component, i.e., poxviruses and adenoviruses. The use of viruses that are non-tropic for mammalians can also circumvent this problem.

As we must assume that the immunizing capacity of recombinant viral cancer vaccines depends on the expression of the tumor antigen in professional APCs, it appears reasonable to transduce DCs ex vivo with the respective viral construct and to use them for immunization (Fig. 4). Indeed, studies in experimental animals demonstrate that such genetically modified DCs can induce protective and therapeutically effective anti-cancer immune responses (Song et al. 1997; Specht et al. 1997).

In the past decade, we have learned that the injection of "naked" plasmid DNA (i.e., DNA without a viral coat), can result in an immune response with a Th1/Tc1 bias. Condon et al. (1996) were the first to show that cutaneous genetic immunization with naked DNA can result in potent, antigen-specific, CTL-mediated protective tumor immunity. Again, direct transfection of DCs appears to be the major mechanism by which naked DNA injections induce anti-tumor immunity (Condon et al. 1996; Porgador et al. 1998). A factor that decisively influences the immunogenicity of naked DNA is the presence or absence of short immunostimulatory sequences (ISS) in the plasmid DNA (Sato et al. 1996). These ISS contain a CpG dinucleotide in a particular base context and induce production of cytokines in various immune cell populations including IL-12 in DCs and IFN-γ in NK cells. For vaccination purposes, CpG-rich ISS are not only administered in the context of naked plasmid DNA but also as an oligodeoxynucleotide adjuvant together with antigen. Such an approach was as effective in the induction of anti-cancer immunity as the vaccination with tumor antigen in complete Freund's adjuvant (Weiner et al. 1997).

Nucleic acids, both RNA and DNA, can also be used to directly transfect DCs ex vivo (Fig. 4) and such DCs are capable of stimulating primary CTL responses in vitro and protective anti-cancer immunity in vivo (Boczkowski et al. 1996). With this approach one can introduce single, well-defined tumor antigen-encoding genes or can use cDNA or mRNA libraries of cancer cells to express the entire unfractionated pool of tumor cell-derived antigens in these APCs. At this moment, it is not yet clear whether RNA- or DNA-transfected DCs will find more application in man. DNA formulations are certainly more stable but, from the viewpoint of safety, the RNA approach bears fewer risks.

9.4 Conclusions

Will the new cancer vaccines quickly revolutionize cancer therapy? We should not be too optimistic because cancer cells can find many ways to escape immune surveillance and attack. These include the loss of tumor-associated antigens and/or HLA, defects in the antigen-processing machinery and, perhaps, the induction of apoptosis in Fas^{+} CTLs by the expression of death ligands (for example, FasL) on melanoma cells. As a consequence, it seems reasonable to use vaccines that activate both the adoptive and the innate immune system, for example, DCs (Fernandez et al. 1999), to administer vaccines together with substances capable of upregulating relevant tumor and histocompatibility antigens, and to find better ways of channeling tumor antigens/peptides into either the MHC class I or class II presentation pathway. Heat shock proteins may be of interest for the latter purpose (Tamura et al. 1997).

A major weakness of many clinical trials with cancer vaccines was the lack of methods to reliably measure the immunogenicity of the vaccine and to correlate the results of specific immunological assays with disease development. This situation has greatly improved in the very recent past by the use of tetrameric MHC molecules to detect an increase in the frequency of peptide-specific T cells (Lee et al. 1999), of ELISPOT assays to determine quality and quantity of cytokines released by tumor cell-reactive T cells and, in the case of melanoma, by the availability of soluble serum markers of the patients' tumor load, for example, the S 100 protein, the melanoma inhibitory activity protein, and 5-S-cysteinyl-dopa.

Using these new methodologies in carefully designed clinical trials will be laborious, time-consuming, and expensive, but crucial for evaluating the efficacy of different types of cancer vaccines.

References

Abdel-Wahab Z, et al (1997) A phase I clinical trial of immunotherapy with interferon-γ-gene-modified autologous melanoma cells. Cancer 80:401

Albert ML, et al (1998) Dendritic cells acquire antigen from apoptotic cells and induce class I-restricted CTLs. Nature 392:86

Belli F, et al (1997) Active immunization of metastatic melanoma patients with interleukin-2-transduced allogeneic melanoma cells: evaluation of efficacy and tolerability. Cancer Immunol Immunother 44:197

Boczkowski D, et al (1996) Dendritic cells pulsed with RNA are potent antigen-presenting cells in vitro and in vivo. J Exp Med 184:465

Boon T, et al (1994) Tumor antigens recognized by T lymphocytes. Annu Rev Immunol 12:337

Borysiewicz LK, et al (1996) A recombinant vaccinia virus encoding human papillomavirus types 16 and 18, E6 and E7 proteins as immunotherapy for cervical cancer. Lancet 347:1523

Bronte V, et al (1997) Antigen expression by dendritic cells correlates with the therapeutic effectiveness of a model recombinant poxvirus tumor vaccine. Proc Natl Acad Sci USA 94:3183

Buschle M, et al (1997) Transloading of tumor antigen-derived peptides into antigen-presenting cells. Proc Natl Acad Sci USA 94:3256

Cella M, Lanzavecchia A (1997) Origin, maturation and antigen presenting function of dendritic cells. Curr Opin Immunol 9:10

Chamberlain RS, et al (1996) Costimulation enhances the active immunotherapy effect of recombinant anti-cancer vaccines. Cancer Res 56:2832

Coiffier B, et al (1998) Rituximab (anti-CD20 monoclonal antibody) for the treatment of patients with relapsing or refractory aggressive lymphoma: a multicenter phase II study. Blood 92:1927

Condon C, et al (1996) DNA-based immunization by in vivo transfection of dendritic cells. Nat Med 2:1122

Crowley M, et al (1990) Dendritic cells are the principal cells in mouse spleen bearing immunogenic fragments of foreign proteins. J Exp Med 172:383

Dalloul A, et al (1992) Interleukin-7 is a growth factor for Sézary lymphoma cells. J Clin Invest 90:1054

Elias EG, et al (1997) Adjuvant immunotherapy in melanoma with irradiated autologous tumor cells and low dose cyclophosphamide. J Surg Oncol 64:17

Fernandez NC, et al (1999) Dendritic cells directly trigger NK cell functions: cross-talk relevant in innate anti-tumor immune responses in vivo. Nat Med 5:405

Gänsbacher B, et al (1990) Interleukin 2 gene transfer into tumor cells abrogates tumorigenicity and induces protective immunity. J Exp Med 172:1217

Golumbek PT, et al (1993) Controlled release, biodegradable cytokine depots: a new approach in cancer vaccine design. Cancer Res 53:5841

Gong J, et al (1997) Induction of antitumor activity by immunization with fusions of dendritic and carcinoma cells. Nat Med 3:558

Hersey P, et al (1990) Evidence that treatment with vaccinia melanoma cell lysates (VMCL) may improve survival of patients with stage II melanoma. Cancer Immunol Immunother 32:173

Hu X, et al (1996) Enhancement of cytolytic T lymphocyte precursor frequency in melanoma patients following immunization with the MAGE-1 peptide-loaded antigen presenting cell-based vaccine. Cancer Res 56:2479

Huang AYC, et al (1994) Role of bone marrow-derived cells in presenting MHC class I-restricted tumor antigens. Science 264:961

Jaeger E, et al (1996) Generation of cytotoxic T-cell responses with synthetic melanoma-associated peptides in vivo: implications for tumor vaccines with melanoma-associated antigens. Int J Cancer 66:162

Johnson RL, et al (1996) Human homolog of patched, a candidate gene for the basal cell nevus syndrome. Science 272:1668

Katsumata M, et al (1995) Prevention of breast tumor development in vivo by downregulation of the $p185^{neu}$ receptor. Nat Med 1:644

Lee PP, et al (1999) Characterization of circulating T cells specific for tumor-associated antigens in melanoma patients. Nat Med 5:677

Mackensen A, et al (1994) Direct evidence to support the immunosurveillance concept in a human regressive melanoma. J Clin Invest 93:1397

Marchand M, et al (1999) Tumor regressions observed in patients with metastatic melanoma treated with an antigenic peptide encoded by gene MAGE-3 and presented by HLA-A1. Int J Cancer 80:219

Matzinger P (1994) Tolerance, danger, and the extended family. Annu Rev Immunol 12:991

McCabe BJ, et al (1995) Minimal determinant expressed by a recombinant vaccinia virus elicits therapeutic antitumor cytolytic T lymphocyte responses. Cancer Res 55:1741

Möller P, et al (1998) Vaccination with IL-7 gene-modified autologous melanoma cells can enhance the anti-melanoma lytic activity in peripheral blood of patients with a good clinical performance status: a clinical phase I study. Br J Cancer 77:1907

Moore PS, Chang Y (1995) Detection of herpesvirus-like DNA sequences in Kaposi's sarcoma in patients with and without HIV infection. N Engl J Med 332:1181

Mueller DL, et al (1989) Clonal expansion versus functional clonal inactivation: a costimulatory signaling pathway determines the outcome of T cell antigen receptor occupancy. Annu Rev Immunol 7:445

Nestle FO, et al (1998) Vaccination of melanoma patients with peptide – or tumor lysate – pulsed dendritic cells. Nat Med 4:328

Pan ZK, et al (1995) A recombinant *Listeria monocytogenes* vaccine expressing a model tumour antigen protects mice against lethal tumour cell challenge and causes regression of established tumours. Nat Med 1:471

Parijs L van, Abbas A (1998) Homeostasis and self-tolerance in the immune system: turning lymphocytes off. Science 280:243

Porgador A, et al (1998) Predominant role for directly transfected dendritic cells in antigen presentation to $CD8^+$ T cells after gene gun immunization. J Exp Med 188:1075

Rescigno M, et al (1998a) Dendritic cell survival and maturation are regulated by different signaling pathways. J Exp Med 188:2175

Rescigno M, et al (1998b) Bacteria-induced neo-biosynthesis, stabilization, and surface expression of functional class I molecules in mouse dendritic cells. Proc Natl Acad Sci USA 95:5229

Riethmüller G, et al (1994) Randomised trial of monoclonal antibody for adjuvant therapy of resected Dukes'C colorectal carcinoma. Lancet 343:737

Rosenberg SA, et al (1988) Use of tumor-infiltrating lymphocytes and interleukin-2 in the immunotherapy of patients with metastatic melanoma. N Engl J Med 319:1676

Rosenberg SA, et al (1998a) Immunologic and therapeutic evaluation of a synthetic peptide vaccine for the treatment of patients with metastatic melanoma. Nat Med 4:321

Rosenberg SA, et al (1998b) Immunizing patients with metastatic melanoma using recombinant adenoviruses encoding MART-1 or gp 100 melanoma antigens. J Natl Cancer Inst 90:1894

Sallusto F, Lanzavecchia A (1994) Efficient presentation of soluble antigen by cultured human dendritic cells is maintained by granulocyte/macrophage colony-stimulating factor plus interleukin 4 and downregulation by tumor necrosis factor α. J Exp Med 179:1109

Sato Y, et al (1996) Immunostimulatory DNA sequences necessary for effective intradermal gene immunization. Science 273:352

Schneeberger A, et al (1999) The tumorigenicity of IL-2 gene-transfected murine M-3D melanoma cells is determined by the magnitude and quality of the host defense reaction: NK cells play a major role. J Immunol 162:6650

Schreiber S, et al (1999) Immunotherapy of metastatic malignant melanoma by a vaccine consisting of autologous interleukin 2-transfected cancer cells: outcome of a phase I study. Hum Gene Ther 10:983

Simons JW, et al (1997) Bioactivity of autologous irradiated renal cell carcinoma vaccines generated by ex vivo granulocyte-macrophage colony-stimulating factor gene transfer. Cancer Res 57:1537

Song W, et al (1997) Dendritic cells genetically modified with an adenovirus vector encoding the cDNA for a model antigen induce protective and therapeutic antitumor immunity. J Exp Med 186:1247
Specht JM, et al (1997) Dendritic cells retrovirally transduced with a model antigen are therapeutically effective against established pulmonary metastases. J Exp Med 186:1213
Staveley-O'Carroll K, et al (1998) Induction of antigen-specific T cell anergy: an early event in the course of tumor progression. Proc Natl Acad Sci USA 95:1178
Steinman RM (1991) The dendritic cell system and its role in immunogenicity. Annu Rev Immunol 9:271
Storey A, et al (1998) Role of a p53 polymorphism in the development of human papilloma virus-associated cancer. Nature 393:229
Tamura Y, et al (1997) Immunotherapy of tumors with autologous tumor-derived heat shock protein preparations. Science 278:117
Toes REM, et al (1996) Enhanced tumor outgrowth after peptide vaccination. Functional deletion of tumor-specific CTL induced by peptide vaccination can lead to the inability to reject tumors. J Immunol 156:3911
Topalian SL, et al (1989) Tumor-specific cytolysis by lymphocytes infiltrating human melanomas. J Immunol 142:3714
Townsend SE, Allison JP (1993) Tumor rejection after direct costimulation of $CD8^+$ T cells by B7-transfected melanoma cells. Science 259:368
Unden AB, et al (1997) Human patched (PTCH) mRNA is overexpressed consistently in tumor cells of both familial and sporadic basal cell carcinoma. Cancer Res 57:2336
Veelken H, et al (1997) A phase-I clinical study of autologous tumor cells plus interleukin-2-gene-transfected allogeneic fibroblasts as a vaccine in patients with cancer. Int J Cancer 70:269
Wang M, et al (1995) Active immunotherapy of cancer with a nonreplicating recombinant fowlpox virus encoding a model tumor-associated antigen. J Immunol 154:4685
Weiner GJ, et al (1997) Immunostimulatory oligodeoxynucleotides containing the CpG motif are effective as immune adjuvants in tumor antigen immunization. Proc Natl Acad Sci USA 94:10833
Wick M, et al (1997) Antigenic cancer cells grow progressively in immune hosts without evidence for T cell exhaustion or systemic anergy. J Exp Med 186:229
Wu TC, et al (1995) Engineering an intracellular pathway for major histocompatibility complex class II presentation of antigens. Proc Natl Acad Sci USA 92:11671

Zatloukal K, et al (1995) Elicitation of a systemic and protective anti-melanoma immune response by an IL-2-based vaccine: assessment of critical cellular and molecular parameters. J Immunol 154:3406
Zitvogel L, et al (1998) Eradication of established murine tumors using a novel cell-free vaccine: dendritic cell-derived exosomes. Nat Med 4:594

10 The Hybrid Cell Vaccination Approach to Cancer Immunotherapy

U. Trefzer, G. Herberth, W. Sterry, P. Walden

10.1 Introduction

A variety of cancer immunotherapy approaches aim at the induction of cytolytic T lymphocytes (CTLs) as main effector cells in anti-tumor immune responses. However, the mere existence of the tumor proves the failure of immune surveillance and points at immune response defects. Antigenicity is prerequisite for specific immunotherapy and has been proven or is suggested by indirect evidence for several tumors. The failure of the immune system to cope with the cancer in these cases must therefore be attributed to immune regulatory problems including immune suppression, the lack of costimulatory support, or, very importantly, the lack of T cell help for the induction of cytolytic T effector cells. In addition, to become instrumental for clinical application, immunotherapy protocols must be suitable for highly individualized treatments and applicable without lengthy preparations.

Following early genetic studies that have demonstrated the need for both permissive MHC class I and MHC class II alleles (Juretic et al. 1985), and the extensive cellular work that has elucidated some of the lymphokines required for the maturation and expansion of cytolytic T effector cells, new concepts for immunotherapy of cancer have been suggested. Both, systemic application of recombinant lymphokines (Maas et al. 1993) and local induction of their genes in gene therapeutically altered cells (Colombo et al. 1992) have been tested. Although many cases of partial and some complete responses have been reported from the corresponding clinical trials, so far, these treatment modalities have disappointed, with response rates between 20% and 30%. The application of recombinant lymphokines is often accompanied by severe side effects, and the gene therapy approaches are still hampered by intolerably long preparation times and low efficiencies.

10.2 The Hybrid Cell Vaccination Concept

Epitope linkage has been shown in our studies on the cellular requirements for the induction of cytolytic T effector cells to be prerequisite for productive T-T cell collaboration (Stuhler and Walden 1993; Borges et al. 1994), i.e., cytolytic precursor and helper T cells have to be activated by the same antigen-presenting cell displaying epitopes for both T cell types on the corresponding MHC class I and class II molecules. Neither the two epitopes nor the responding T cells need to be related. The implications of this concept are, first, only cells expressing both MHC class I and II molecules can induce cytolytic T cell responses (Stuhler and Walden 1993; Grabbe et al. 1995; Young and Inaba 1996), second, cognate antigens must be presented for both T cell and MHC types (Mitchison and O'Malley 1987; Mitchison 1990; Stuhler and Walden 1993; Bennet et al. 1998; Ridge et al. 1998; Schoenberger et al. 1998), and, third, there must be T cells with the corresponding specificities in the T cell receptor repertoire of the response-competent individual. The hybrid cell vaccination approach to cancer immunotherapy was developed to meet the above requirements (Stuhler and Walden 1994). While

MHC class I-restricted CTL epitopes are needed for the anti-tumor immunity and, thus, are determined by the tumor's antigenicity, MHC class II-restricted epitopes can be chosen freely to avoid limitations in helper T cell frequencies and and their ability to be activated (Borges et al. 1994; Stuhler and Walden 1994). The notoriously vigorous allo MHC-specific T cell responses are among the most dependable immune responses and therefore a well-suited source for T cell help. Combining all these considerations, fusion of autogeneic tumor cells and allogeneic MHC class II-bearing cells might render antigenic tumor cells immunogenic. In addition, the fusion partner may also complement potential costimulatory deficiencies of the tumor cell and, thus, enhance the efficacy of the immunogen (Chen et al. 1992; Guo et al. 1994). This concept has been tested in animal model experiments for thymoma (Stuhler and Walden 1994) and hepatocarcinoma (Guo et al. 1994). The animals could be cured efficiently from the transplanted tumors and acquired a long-lasting immunity to subsequent challenges with the same tumor. Moreover, in these animal systems it could be shown that mixing allogeneic B cells and tumor cells was not effective (Guo et al. 1994; Stuhler and Walden 1994). These experiments have provided in vivo evidence for the epitope linkage requirement for helper T cell-dependent CTL induction as well as for the involvement of the B7-CD28 costimulatory pathway.

10.3 The First Clinical Phase I Trial with Hybrid Cell Vaccination

We have conducted the first clinical application of the hybrid cell vaccination approach in 16 patients with metastatic melanoma (Trefzer et al. 2000). These patients were well advanced in their disease, had involvement of more than one organ, and had previously failed other treatment modalities such as chemotherapy and/or immunotherapy with cytokines (IFN-α, IL-2). The primary objective of this clinical phase I trial was to test efficacy, toxicity, and feasibility of this treatment.

Complete absence of delayed-type hypersensitivity (DTH) responses to recall antigens was seen in 6 of the 15 patients and had to be taken as indication of a severely compromised immune system and a lack of cellular immune response capacity. None of the participants in the study mounted an anti-tumor DTH response prior to vaccination.

10.4 Tumor and B Cells

Single tumor cell suspensions were prepared from the tumors immediately after surgical excision. After removing fat and fibrous tissue the tumors were minced and the pieces incubated under agitation for 1–2 h at 37°C in RPMI 1640 medium containing collagenase type VIII and DNase type IV (26 μg/ml). The cell suspension was then passed through a tissue sieve and washed 3 times with PBS at room temperature and centrifuged at 800 *g*. The viability of the cells was determined by dye exclusion and usually exceeded 90%.

To generate tumor cell lines, the cells were grown for the first 2 days in RPMI 1640 medium with FCS (10%), hydrocortisone (0.5 μg/ml; Sigma), insulin (10 μg/ml; Boehringer Biochemicals), PMA (10 ng/ml; Sigma), non-essential amino acids (1%), penicillin (50 IU/ml), streptomycin (50 μg/ml), bFGF (1 ng/ml), EGF (5 ng/ml), and l-glutamine (2 mM; all Gibco) at 37°C and 5% CO_2 in plastic dishes. After 2 days the medium was changed to RPMI 1640 with FCS (10%), glutamine (2 mM), and antibiotics. Tumor cell lines were established for 13/15 patients.

Allogeneic PBMCs were prepared from buffy coats obtained from healthy donors by Ficoll gradient centrifugation. To enrich for activated B cells carrying MHC class II and costimulatory molecules, the PBMCs were cultured for 3 days in RPMI 1640 medium containing FCS (10%) with LPS (50 μg/ml), rhIL-4 (100 U/ml), and rhIL-6 (100 U/ml). The cells were then double stained with the PE-labeled B cell-specific anti-CD19 monoclonal antibody and FITC-labeled monoclonal antibodies with specificities for B7–1, B7–2, ICAM-1, MHC class I, or MHC class II. The monoclonal antibodies were incubated with the cell suspensions for 45 min at 4°C, washed 3 times, fixed with 1% paraformaldehyde for 5 min, and analyzed by flow cytometry (FACScan; Becton Dickinson). The activation of the B cells was confirmed by the high levels of MHC class I, MHC class II, ICAM-1, and B7–1 and/or B7–2 expression.

10.5 Production of the Hybrid Cell Vaccine

Fusion of the tumor cells with the stimulated B cells is achieved by electric fusion (Stuhler and Walden 1994; Trefzer et al. 2000). Allogeneic PBMCs from healthy donors activated for 3 days in culture medium containing LPS (50 µg/ml), rhIL-4 (100 U/ml), and rhIL-6 (100 U/ml) and single cell suspensions of melanoma cells are washed separately 4 times in sterile glucose solution (5%). Immediately before fusion the cells are mixed at a ratio of 1:1 to yield a density of 2×10^7 cells/ml. The cells are dielectrophoretically aligned by an inhomogenous electric field in an electroporation chamber and fused with a single pulse of 1000 V/cm, at a 25-µF setting of a gene pulser. The cell viability after the fusion is determined by dye exclusion (usually above 80%). The fused cells are washed, irradiated at 200 Gy, and injected subcutaneously in 5 ml glucose (5%). The fusion efficiencies were assessed in independent tests whereby the partner cells are stained separately with different fluorescent membrane dyes and subjected to the above electrofusion procedure. The efficiencies of heterogeneous fusions are usually 20%–30% as indicated by the occurrence of doubly stained cells detected by flow cytometry or fluorescence microscopy. All the cell materials used for the production of the vaccines were tested for the absence of viruses.

10.6 Treatment

The patients received at least two subcutaneous injections (range 2–7) of the vaccine with at least 3×10^6 tumor cells (range $3–10 \times 10^6$) in 1.5–5 ml glucose (5%). The inoculums were divided equally between two separate sites, usually the lower abdomen and the upper thigh. Upper arms were used if lymph nodes in the inguinal area had been removed. The tumor response was assessed 3 months after the first injection and thereafter every 3 months.

10.7 Toxicity and Side Effects

The hybrid cell vaccination caused only minor side effects: 5 patients experienced no side effects apart from the local erythema or induration at the site of inoculation which was seen in 14 patients. In three cases these reactions were accompanied by local pain. Three patients responded with chills, 8 with low grade fever, and 6 with bouts of sweating. These reactions were strongest within 24 h after vaccination and waned within 72 h with the exception of one case where profuse sweating throughout the entire course of the treatment was observed. One patient developed a localized eczema on his left thigh in an area with subcutaneous tumor lesions and scar tissue. Two patients developed small areas of vitiligo which was restricted in 1 patient to the site of cutaneous melanoma metastases. No other sign of vaccination-induced autoimmune reactions was observed. All together, there was no correlation between the strength of the side effects and clinical outcome of the therapy attempt.

10.8 Clinical Outcome and DTH Reaction

None of the 6 patients who failed to mount DTH responses to recall antigens showed any sign of clinical anti-tumor response. With regard to the clinical responses, 7 of the 15 patients responded to the hybrid cell vaccination, 2 with complete responses and 5 with stable disease for 4, 6, 14, 24, and 30 months, respectively. One of the patients with complete response experienced a partial response of a large tumor nodule in the pelvis. Vaccination was continued for a prolonged period of time on a monthly basis. After 32 months, staging examinations showed a complete remission of the tumor which was replaced by scar tissue (Table 1). In 4 out of 16 patients, a DTH reaction against autologous tumor cells was observed after vaccination. This correlated in 3 patients with a clinical response to the treatment. However, in the other three responders, no DTH reaction was evident.

Table 1. Clinical characteristics

Patient number	Previous therapy	Number of organs involved	Mérieux test[a]	Number of injections	Cell line used	Response
1	S,C	2	Negative	6	+	PD
2	S	2	+	5	–	PD
3	S	3	+	3	–	SD
4	S,CI	3	+	4	+	PD
5	S,C,I	4	Negative	4	+	PD
6	S,C,I	2	+	3	–	SD
7	S,C	3	Negative	7	+	PD
8	S,C	3	Negative	4	+	PD
9	S,CI	1	+	2	–	CR
10	S,R	2	+	4	+	SD
11	S	2	Negative	2	–	PD
12	S,R,C,I	6	+	3	–	SD
13	S,R	4	+	3	+	PD
14	S,C,I	2	Negative	5	+	PD
15	S,C,CI,I	1	+	5	–	SD
16	S,C	1	+	19	–	CR

S, surgery; R, radiotherapy; C, chemotherapy; CI, chemoimmunotherapy; I, immunotherapy; CR, complete remission; SD, stable disease; PD progressive disease

[a]Mérieux test + if at least one reaction against one of the recall antigens was positive: proteus, tetanus toxin, tuberculin, staphylococcus, trichophyton, candida, diphtheria

10.9 Use of Dendritic Cells (DCs) for Hybrid Cell Vaccination

This approach is undertaken since DCs should be at least equal in the antigen-presenting capacity when compared to B cells. This approach has been facilitated by recently described methods for generating large numbers of DCs from the peripheral blood of donors.

DCs are generated by adding GM-CSF (50 ng/ml) and IL-4 (1000 U/ml) to monocytes obtained after adherence of PBMCs collected by leukapheresis of healthy donors. Single cell suspensions of autologous melanoma metastasis are mixed with DCs at a ratio of 1:3

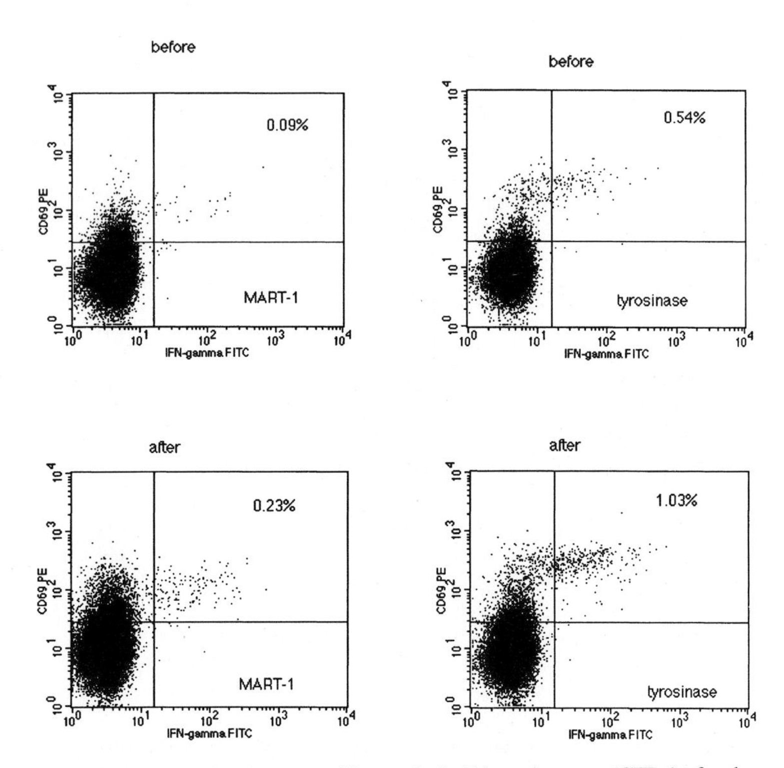

Fig. 1. Induction of antigen-specific cytolytic T lymphocytes (CTLs) after hybrid cell vaccination. Before vaccination, MelanA/Mart-1 (*top left*)- or tyrosinase (*top right*)-specific CTLs can be detected in the peripheral blood of patients. Twenty-four hours after vaccination, the frequencies of both MelanA/Mart-1 (*bottom left*)- and tyrosinase (*bottom right*)-specific CTLs are significantly higher

and fused by electrofusion. The presence of fused cells is verified by double staining of hybrid cells with DC and tumor cell markers. The hybrid cells express high levels of class I and II, ICAM-1, B7–1, and B7–2. For evaluating the in vivo efficacy we have initiated a clinical trial where patients with metastatic melanoma receive an irradiated hybrid cell vaccine on a monthly basis. Again, this treatment is so far well tolerated. This trial is accompanied by analyses where we ask the

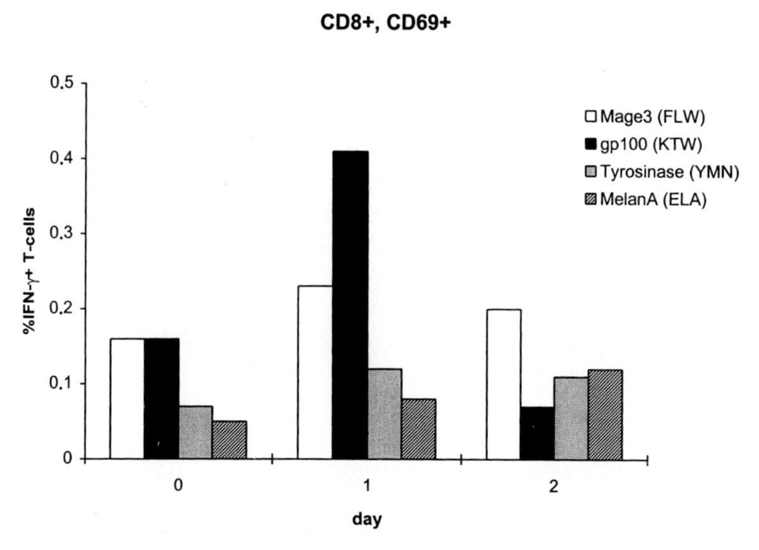

Fig. 2. Induction of CTLs in a vaccine-treated patient. All analyzed CTL frequencies (specific for MAGE-3, gp100, tyrosinase, or MelanA/Mart-1) are higher 1 day after treatment and subsequently decline after 2 days

question, whether this vaccination approach is capable of inducing antigen-specific immune responses.

The activation of CTLs is the most critical component in therapeutic vaccine studies of human cancer (Lethe et al. 1992; Boon et al. 1994). Therefore, the identification of antigen specific T lymphocytes is central for the analysis of the intended immune response in various immunotherapies of cancer and the induction of CD8+ T cells in vivo can be considered an important end-point in vaccination studies (Altman et al. 1996). Methods for analysis of T cell function have traditionally relied upon measurements of proliferation, ^{51}Cr release assays, limiting dilution assays, or cytokine expression in bulk cultures of PBMCs (measurement of secreted cytokines and cytokine mRNA from non-clonal populations), as well as ELISPOT techniques, that allow determination of the number of cells secreting a single cytokine. All of these methods are time consuming and labor intensive. We have established a flow cytometric technique for the detection of antigen-specific T cells in peripheral blood and PBMCs which is based on intracellular IFNγ expression after specific peptide stimulation. As an example we demonstrate the findings in two patients receiving hybrid cells. In patient number one a significant increase of antigen-specific CTLs could be detected in the peripheral blood 1 day after vaccination. Both MelanA/Mart-1 and tyrosinase-specific CTLs were found to be significantly increased (Fig. 1). These cells are activated, IFNγ-producing, $CD8^+$, $CD69^+$ CTLs. In patient number two similar observations were made. The increase of MAGE-3-, gp100-, tyrosinase-, and MelanA/Mart-1-specific CTLs 1 day after vaccination therapy was followed by a decline to pretreatment levels after 2 days by some of the antigen-specific CTLs (Fig. 2).

10.10 Conclusions

The hybrid cell vaccination approach to cancer immunotherapy has been designed:

- to recruit and activate T cell help for the induction of tumor-specific cytotoxic T cells
- to correct defects in costimulatory signaling
- to utilize a large variety of usually unidentified tumor-associated antigens
- for individualized therapy that can be applied instantly without long preparations.

As a result of our experience with this approach to cancer immunotherapy it appears that hybrid cell vaccination is a safe procedure which is well acceptable for the patients. The preparation of the vaccine is relatively simple and time efficient. In some cases of an easily accessible tumor nodule for the hybrid cell production and a relatively good constitution, the vaccines can be given at out-patient clinics after the initial immunization. Hence, this approach is suited for individualized treatment and may become a cost-effective therapy.

The low number of patients and their clinical profiles restrict firm statements on the efficacy of hybrid cell vaccination for cancer immunotherapy which have to await large-scale randomized phase III trials. The initial observations are promising, however, and suggest that the trials should be continued and extended to other tumor types. In this regard, a clinical trial in patients with metastatic renal cell carcinoma (Kugler et al. 1998) has revealed similar results as in our trial. While this treatment was also well tolerated by the patients, tumor regressions were seen in some of the patients.

References

Altman JD, Moss PAH, Goulder PJR, et al (1996) Phenotypic analysis of antigen-specific T lymphocytes. Science 274:94–96

Bennet SRM, Carbone FR, Karamalis F, Flavell RA, Miller JFAP, Heath WR (1998) Help for cytotoxic-T-cell responses is mediated by CD40 signalling. Nature 393:478–480

Boon T, Cerottini JC, Eynde B van den, Bruggen P van der, Pel A van (1994) Tumor antigens recognized by T lymphocytes. Annu Rev Immunol 12:337–365

Borges E, Wiesmüller KH, Jung G, et al (1994) Efficacy of synthetic vaccines in the induction of cytotoxic T lymphocyte responses. J Immunol Methods 173:253–263

Chen LP, Ashe S, Brady WA, et al (1992) Costimulation of anti-tumor immunity by the B7 counterreceptor for the T lymphocyte molecule CD28 and CTLA4. Cell 71:1093–1102

Colombo MP, Mattei S, Parmiani G (1992) Cytokine gene transfer in tumor cells as an approach to antitumor therapy. Int J Lab Res 21:278–282

Grabbe S, Beissert S, Schwarz T, et al (1995) Dendritic cells as initiators of tumor immune responses: a possible strategy for tumor immunotherapy? Immunol Today 16:117–121

Guo YJ, Wu M, Chen H, et al (1994) Effective tumor vaccine generated by fusion of hepatoma cells with activated B cells. Science 263:518–520

Juretic A, Malenica B, Juretic E, et al (1985) Helper effects required during in vivo priming for a cytolytic response to the H-Y antigen in nonresponder mice. J Immunol 134:1408–1410

Kugler A, Seseke F, Thelen P, Kallerhoff M, Muller GA, Stuhler G, Muller C, Ringert RH (1998) Autologous and allogenic hybrid cell vaccine in patients with metastatic renal cell carcinoma. Br J Urol 82:487–493

Lethe B, Eynde B van den, Pel A van, Corradin G, Boon T (1992) Mouse tumor rejection antigens P815 A and P815 B: two epitopes carried by a single peptide. Eur J Immunol 22:2283–2288

Maas RA, Dullens HFH, DenOtter W (1993) Interleukin-2 in cancer treatment – disappointing or (still) promising. Cancer Immunol Immunther 66:141–148

Mitchison NA (1990) An exact comparison between the efficiency of two-and three-cell-type clusters mediating helper activity. Eur J Immunol 20:699–702

Mitchison NA, O'Malley C (1987) Three-cell-type clusters of T cells with antigen-presenting cells best explain the epitope linkage and noncognate requirements of the in vivo cytolytic response. Eur J Immunol 17:1479–1483

Ridge JP, Di Rossa F, Matzinger P (1998) A conditioned dendritic cell can be a temporal bridge between a CD4+ T-helper and a T-killer cell. Nature 393:474–478

Schoenberger SP, Toes REM, Voort EIH van der, Offringa R, Melief CJM (1998) T-cell help for cytotoxic T lymphocytes is mediated by CD40-CD40L interactions. Nature 393:480–483

Stuhler G, Walden P (1993) Collaboration of helper and cytotoxic T lymphocytes. Eur J Immunol 23:2279–2286

Stuhler G, Walden P (1994) Recruitment of helper T-cells for induction of tumor rejection by cytolytic T lymphocytes. Cancer Immunol Immunother 39:342–345

Trefzer U, Weingart G, Chen YW, Audring H, Winter P, Guo YJ, Sterry W, Walden P (2000) Hybrid cell vaccination for cancer immunotherapy: first clinical trial with metastatic melanoma. Int J Cancer 85:618–626

Young JW, Inaba K (1996) Dendritic cells as adjuvants for class I major histocompatibility complex-restricted antitumor immunity. J Exp Med 183:7–11

11 T Cell Receptor Peptides for the Vaccination Therapy of Multiple Sclerosis

S. Brocke

11.1 Introduction

Current options for treatment of diseases with a presumed autoimmune etiology such as multiple sclerosis (MS) are far from satisfactory. Most treatments are associated with undesirable side effects due to toxicity and lack of immunological specificity. For these reasons, one of the primary goals in the development of immunotherapies has been to achieve selective inactivation of disease-inducing lymphocytes in the absence of general immunosuppression. It appears that the peripheral immune system can be divided into two compartments: one in which positive responses are initiated and another one where tolerance is induced (Mitchison 1998). Augmenting of tolerance by immunoregulation ranks prominently among the various approaches studied for the treatment of autoimmune disorders. Strategies in this direction include induction of specific immunological tolerance via anergizing, deletion, or suppression of autoreactive clones (Van Paris et al. 1998). The role of some of these mechanisms in maintaining peripheral tolerance in vivo is still very much a matter of debate.

One approach that has received much attention recently is T cell vaccination, originally proposed by Cohen and Ben-Nun (Ben-Nun and Cohen 1981; Ben-Nun et al. 1981). Their initial observation was that radiation-inactivated encephalitogenic T cells could be used as preventive vaccines against the induction of experimental autoimmune encephalomyelitis (EAE). Based on these early findings, other means of inactivating potentially autoreactive T cells, including high pressure, alkylating agents such as mitomycin C or chemical crosslinkers, were used in order to successfully attenuate myelin basic protein (MBP)-specific T cells. Upon adoptive transfer, these cells conferred resistance to subsequent active EAE induction in naïve recipients. Successful T cell vaccination therapy was subsequently demonstrated in animal models of arthritis (Lider et al. 1987; Kumar et al. 1997), lupus (Ben-Yehuda et al. 1996), and type I diabetes (Elias et al. 1999). Encouraging results in animal models using vaccines based on the pathogenic T cell have prompted the design of novel and selective immune-based experimental therapies for human autoimmune disease.

Although the protective effect of T cell vaccination in vivo has been extensively documented, the precise mechanism for this effect has remained unclear. Since initial studies, much evidence has been accumulated in support of the function of regulatory T cell antigen receptor (TCR)-specific T lymphocyte networks (Lider et al. 1988). Current concepts of the pathophysiology of autoimmune disease point to the crucial role of pathogenic Th1 T cells, autoantigenic epitopes, major histocompatibility complex (MHC) molecules, and possibly regulatory T cell populations in the disease process. It is widely accepted that Th1 cells, critical for cell-mediated immunity by their production of IL-2, IFN-γ, TNF-α, and lymphotoxin are involved in the immunopathology of organ-specific autoimmune disease (O'Garra et al. 1997) A role as regulators has been suggested for Th2 cells (O'Garra et al. 1997) and cells producing TGF-β, recently characterized as Th3 and Tr (regulatory) $CD4^+$ T cells. Results from vaccination experiments of rats with a low, subencephalitogenic dose of a pathogenic T cell clone suggested that T cell vaccination induced resistance to autoimmune disease by activating an anti-idiotypic network of T cell responses (Lider et al. 1988). These T cells could function as regulatory Th2, Th3, or Tr cells (Cohen 1991; Vandenbark et al. 1996a,b). It was suggested that this phenomenon is driven by determinants (idiotopes) on the antigen receptors of the pathogenic Th1 cells used as a vaccine. For this reason, the research on protective T cell vaccination focused on the identification of TCR peptides that serve as antigenic epitopes of an immunoregulatory potentially anti-idiotypic T cell response.

11.2 Antigen Recognition by the T Cell Antigen Receptor

The T cells of the immune system recognize antigens as immunogenic peptides non-covalently bound to MHC class I and class II molecules expressed on the surface of antigen-presenting cells. The antigen/MHC binding region of the TCR is a heterodimer composed of α and β transmembrane glycoprotein chains. The extracellular portion of each chain contains two domains, resembling immunoglobulin variable and constant domains. The V-like domains comprise V-, D-, and J-elements in the β chain and V- and J-elements in the α chain and are encoded by corresponding rearranged T cell receptor gene segments. The possible number of different T cell receptor gene rearrangements is estimated to be over 10^{15}, thus ensuring potential recognition of virtually any antigen. The third complementarity-determining regions (CDR)3 of the TCR α and β chains, to which the D and J gene segments contribute, form the center of the antigen-binding site on T cell receptors. In contrast, the periphery of the binding site consists of CDR1 and CDR2 loops, which are encoded within the germline V gene segments for the α and β chains. For this reason, all the diversity in T cell receptors is generated during rearrangement and is consequently focused on the "clonotypic" CDR3 regions.

11.3 TCR/TCR Recognition

Evidence supporting the action of regulatory networks has been found in many in vitro and in vivo studies. Initially, it was discovered that vaccination with attenuated encephalitogenic, arthritigenic, or diabeto-

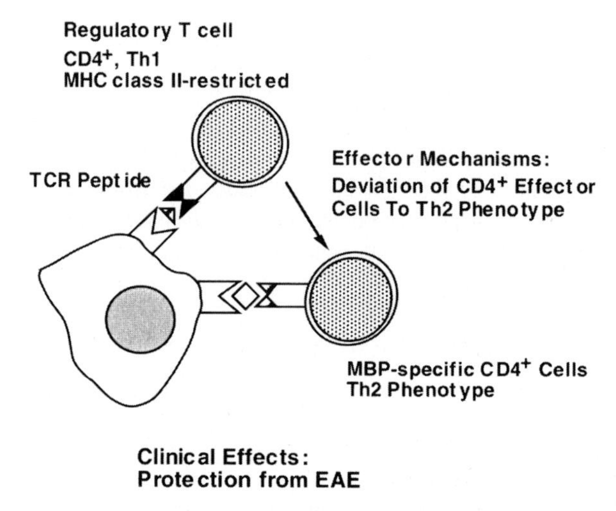

Fig. 1. Regulatory T cell network in B10.PL mice. Th1 $CD4^+$ cells polarize myelin basic protein (*MBP*)-specific effector cells into Th2

genic T cells can prevent induction of EAE, experimental arthritis, or insulin-dependent diabetes mellitus in animal models (Cohen 1991). In some of these models, anti-clonotypic T cell responses were observed, presumably against immunogenic regions of the TCR.

One of the striking features of experimental autoimmunity is the tendency of autoreactive T cells to utilize a limited V gene repertoire in forming functional TCRs specific for defined autoantigens (Heber-Katz and Acha-Orbea 1989). In many murine autoimmune disease models and in the Lewis rat, autoreactive T cells utilizing the TCR Vβ8.2 gene segment play a major role. Major efforts have been directed at the characterization of immunogenic and therapeutic epitopes on the TCR of the restricted autoreactive T cell population (Vandenbark et al. 1996a,b).

Kumar and colleagues were able to demonstrate that spontaneous recovery from EAE in B10.PL mice is associated with the induction and expansion of regulatory $CD4^+$ T cells that recognize a framework (FR)3 region determinant within the TCR Vβ8.2 chain (Kumar and Sercarz 1993, 1998; Kumar et al. 1995, 1997). These cells appear to be involved

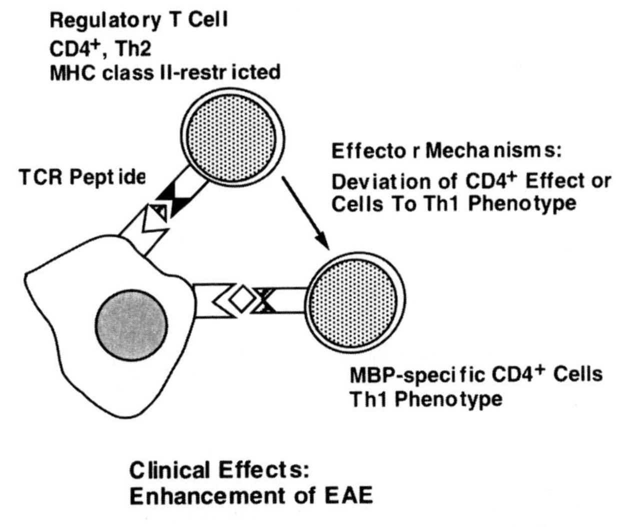

Fig. 2. Regulatory T cell network in B10.PL mice. Th2 $CD4^+$ cells polarize myelin basic protein (*MBP*)-specific effector cells into Th1

in the protection and recovery from EAE since inactivation results in increased severity of disease and poor or delayed recovery (Kumar et al. 1996). In addition, regulatory T cells specific for the same FR3 region of the Vβ8.2 chain are involved in the control of collagen II-induced arthritis (Kumar et al. 1997). It was subsequently shown that the mechanism of regulatory T cell-mediated protection from disease was based on the ability of anti-idiotypic cells to influence the polarization of T cell responses. Thus, the cytokine profile of TCR peptide-specific regulatory T cells influences the cytokine profile of MBP-specific effector cells during priming. Action of Th1-type regulatory T cells results in the eventual dominance by a Th2-deviated population of $CD4^+$ effector cells (Fig. 1). In contrast, in conditions favoring the differentiation of Th2-type $CD4^+$ regulatory cells, the frequency of MBP-specific Th1-type effector T cells increases (Fig. 2). The authors conclude that induction or protection from autoimmune disease depends on the cytokine secretion profile of anti-idiotypic T cells in vivo (Kumar and Sercarz 1998).

Table 1. T cell receptor determinants implicated in regulatory networks

Strain and species	Region	Clinical effect	Reference
PL/J mouse	CDR2, TCR Vβ8.2, p39–61	Not tested	Gaur et al. (1993)
B10.PL mouse	FR3, TCR Vβ8.2, p76–101	Protection from EAE	Kumar and Sercarz (1993, 1998); Kumar et al. (1995, 1996, 1997)
DBA/1LacJ	FR3, TCR Vβ8.2, p76–101	Protection from CIA	Kumar et al. (1997)
Lewis rat	CDR2, TCR Vβ8.2. p39–59	Protection No effect Exacerbation	Vandenbark et al. (1989, 1996a,b); Desquenne-Clark et al. (1991)
Human	CDR2, TCR Vβ5.2, Vβ6.1, 38–58	Not established	Vandenbark et al (1996a,b)
Human	FR1, CDR3, $TCR_{HWBP\text{-}3}$ α chain, FR1, CDR1, CDR2, $TCR_{HWBP\text{-}3}$ β chain	Not established	Zipp et al. (1998)

CDR, complementarity-determining region; TCR, T cell antigen receptor; EAE, experimental autoimmune encephalomyelitis; CIA, collagen-induced arthritis

Several groups successfully demonstrated complete protection from developing clinical signs of EAE using the rat TCR Vβ8.2–39–59 CDR2 peptide (Vandenbark et al. 1989, 1996b). Others, however, failed to detect clinical benefits or observed enhanced EAE severity after preimmunization with the same CDR2 peptide (Desquenne-Clark et al. 1991; Vandenbark et al. 1996b). In the PL/J mouse, a slightly longer TCR Vβ8.2–39–61 CDR2 peptide induced $CD8^+$-dependent clonal anergy in antigen-specific TCR Vβ8.2 T cells (Gaur et al. 1993). Howell and colleagues (1989) showed protective activity against EAE using synthetic peptides corresponding to CDR3 determinants of the VDJ region of a TCR β chain conserved among encephalitogenic T cells. Some of the studies demonstrating protective effects of synthetic TCR peptide vaccines in EAE or other autoimmune diseases are summarized in Table 1.

11.4 Immune Responses Directed Against the T Cell Receptor in Humans

While the action of regulatory T cell networks could easily be detected in defined animal models of autoimmune disease, it was more of a challenge to demonstrate similar circuits operating in other models of EAE or in humans. It became clear that in some mouse strains such as SJL or in humans, T cell responses to various myelin antigens are more complex. Both antigenic determinants on MBP, proteolipid protein or myelin oligodendrocyte glycoprotein are variable, and the TCR usage of responding cells can be quite heterogenous. Several studies showed that human TCR α and β chains contain multiple determinants that are processed and presented by antigen-presenting cells and elicit a specific T cell response. Using two synthetic peptides derived from a particular TCR Vβ chain, it was shown that it is possible to select TCR peptide-specific, MHC class II-restricted CD4$^+$ T cells from the normal, unprimed human repertoire (Saruhan-Direskeneli et al. 1993). In the meantime, a detailed determinant analysis of the human anti-T cell response was reported (Bourdette et al. 1994, 1998; Chou et al. 1994; Vandenbark et al 1996a). In a small cohort of MS patients, circulating T cells specific for human Vβ5.2 (38–58) and Vβ6.1 (38–58) could clearly be demonstrated and in some cases boosted with the appropriate peptides (Vandenbark et al. 1996a,b; Bourdette et al. 1998). Immunization with either natural or substituted (Tyr49Thr) peptides corresponding to TCR Vβ5.2 (38–58) reduced in some patients the number of MBP-reactive T cells (Vandenbark et al. 1996a). Injections of altered TCR Vβ5.2 (38–58) peptide were more effective at breaking tolerance to self-TCR. The peptide-specific T cells were polarized toward a Th2 cytokine profile characterized by high IL-10 production and low IFN-γ. In vitro, supernatants from these cells as well as IL-10 were shown to be capable of inhibiting MBP-specific T cell proliferation. These findings suggest that it is possible to generate Th2 cytokine-producing bystander suppressor cells specific for TCR Vβ5.2 (38–58) peptides or their derivatives (Fig. 3).

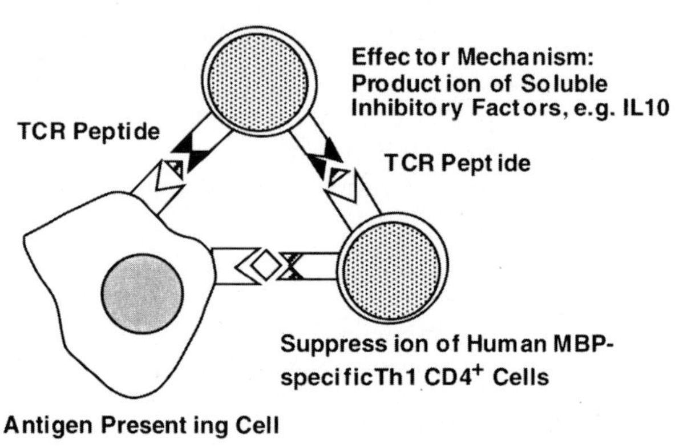

Fig. 3. Presentation of T cell antigen receptor (*TCR*) peptides by human $CD4^+$ T cells and antigen-presenting cells to a regulatory $CD4^+$ T cell. Soluble factors, including Th2 cytokines, suppress myelin basic protein (*MBP*)-specific effector cells

Another approach in humans has been the induction of clonotypic regulatory networks by immunization with bulk preparations of irradiated MBP-specific T cells (Zhang et al. 1995). This type of immunization induced predominantly $CD8^+$ regulatory T cells capable of lysing the immunizing clones in a clonotype-specific manner (Fig. 4).

In a detailed study, Zipp and coworkers (1998) successfully identified immunogenic TCR epitopes that map to the FR1 and CDR3 regions of the Vα chain and FR1, CDR1, and CDR2 regions of the Vβ chain expressed by a T cell clone with specificity for MBP. This study demonstrates that multiple TCR α and β chain epitopes can exist on a single T cell clone and indicates the need to determine the potentially marked diversity of the anti-TCR response in each patient before vaccination therapies can be applied. T cells recognized the identified TCR epitopes in the context of at least three different HLA-DR molecules. However, most TCR peptide-specific T cells in this study were selected using synthetic peptides and did not respond to full-length TCR V chains. This

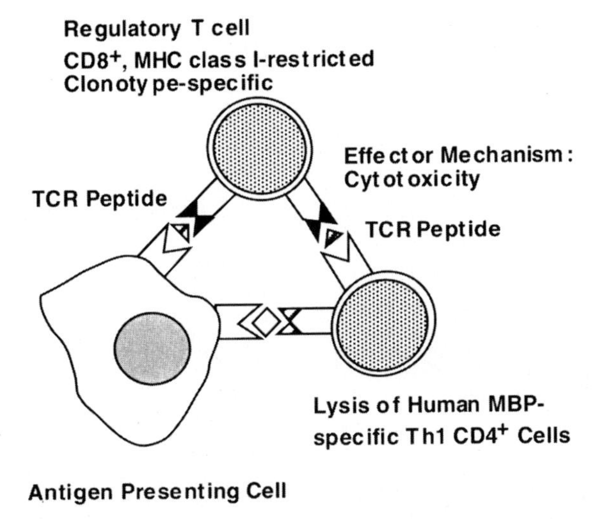

Fig. 4. Presentation of T cell antigen receptor (*TCR*) peptides by human CD4$^+$ T cells and antigen-presenting cells to a regulatory CD8$^+$ T cell. CD8$^+$ T cells specifically lyse myelin basic protein *(MBP)*-reactive effector cells

observation raises the possibility that a fraction of TCR epitopes determined by measuring immune responses against synthetic peptides represent cryptic determinants (Zipp et al. 1998). Activated human T cells express both MHC class I and class II molecules. Next to dendritic cells and macrophages, T cell-bound MHC class II may be a potential site for the presentation of clonotypic TCR fragments. These presented TCR epitopes could be the target structures for TCR-specific regulatory cells, which suppress via inhibitory cytokines or lyse autoreactive T cells (Fig. 5).

11.5 Conclusions

The presence of antigenic epitopes on myelin antigen-specific T cell receptors has been clearly established in various species, including humans. Clinical and immunological observations in a group of MS patients undergoing TCR peptide vaccination provide encouragement

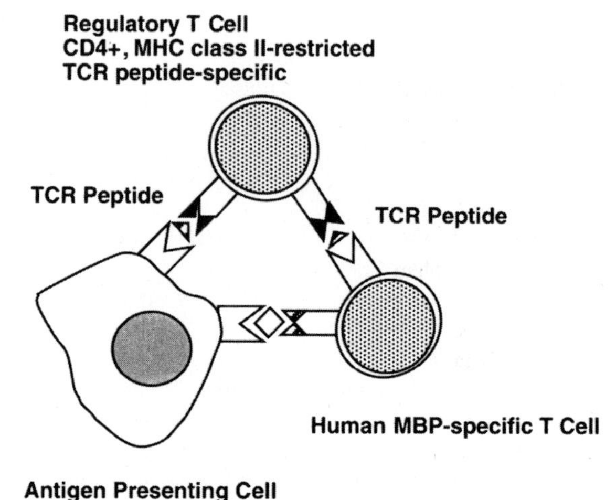

Fig. 5. Presentation of T cell antigen receptor (*TCR*) peptides by human CD4$^+$ T cells and antigen-presenting cells to a TCR-specific T cell

for further patient studies examining the potential of therapeutic TCR vaccination in autoimmune diseases (Antel et al.1996). It appears that in contrast to some animal models, human anti-TCR immune responses may be characterized by a varying degree of diversity, both with respect to epitopes on defined TCRs as well as TCR usage in response to autoantigens. A detailed study of autoantigen-specific TCRs and their epitopes that are recognized by T cells of individual patients may therefore be of critical importance in selecting candidate vaccines. An alternative approach may be vaccination with DNA encoding full-length TCR V regions (Waisman et al. 1996).

Acknowledgements. The work in the author's laboratory was supported by a grant from the Israel Science Foundation, administered by the Israel Academy of Sciences. S.B. is incumbent of the Dr.-Norman-Joels-Lectureship in Life Sciences and Medicine.

References

Antel JP, Becher B, Owens T (1996) Immunotherapy for multiple sclerosis: from theory to practice. Nat Med 2:1074–1075

Ben-Nun A, Cohen IR (1981) Vaccination against autoimmune encephalomyelitis (EAE): attenuated autoimmune T lymphocytes confer resistance to induction of active EAE but not to EAE mediated by the intact T lymphocyte line. Eur J Immunol 11:949–952

Ben-Nun A, Wekerle H, Cohen IR (1981) Vaccination against autoimmune encephalomyelitis with T-lymphocyte line cells reactive against myelin basic protein. Nature 292:60–61

Ben-Yehuda A, Bar-Tana R, Livoff A, Ron N, Cohen IR, Naparstek, Y (1996) Lymph node cell vaccination against the lupus syndrome of MRL/lpr/lpr mice. Lupus 5:323–326

Bourdette DN, Witham RH, Chou YK, Morrison WJ, Atherton J, Kenny C, Liefeld D, Hashim GA, Offner H, Vandenbark AA (1994) Immunity to TCR peptides in multiple sclerosis. I. Successful immunization of patients with synthetic V beta 5.2 and V beta 6.1 CDR2 peptides. J Immunol 152:2510–2519

Bourdette DN, Chou YK, Witham RH, Buckner J, Kwon HJ, Nepom GT, Buenafe A, Cooper SA, Allegretta M, Hashim GA, Offner H, Vandenbark AA (1998) Immunity to T cell receptor peptides in multiple sclerosis. III. Preferential immunogenicity of complementarity-determining region 2 peptides from disease-associated T cell receptor BV genes. J Immunol 161:1034–1044

Chou YK, Morrison WJ, Weinberg AD, Dedrick R, Witham R, Bourdette DN, Hashim GA, Offner H, Vandenbark AA (1994) Immunity to TCR peptides in multiple eclerosis. II. T cell recognition of Vbeta 5.2 and Vbeta 6.1 CDR2 peptides. J Immunol 152:2520–2529

Cohen IR (1991) T cell vaccination in immunological disease. J Intern Med 230:471–477

Desquenne-Clark L, Esch TR, Otvos L, Heber-Katz E (1991) T cell receptor peptide immunization leads to enhanced and chronic experimental allergic encephalomyelitis. Proc Natl Acad Sci USA 88:7219–7223

Elias D, Tikochinsky Y, Frankel G, Cohen IR (1999) Regulation of NOD mouse autoimmune diabetes by T cells that recognize a TCR CDR3 peptide. Int Immunol 11:957–966

Gaur A, Ruberti G, Haspel R, Mayer JP, Fathman CG (1993) Requirement for CD8+ cells in T cell receptor peptide-induced clonal unresponsiveness. Science 259:91–94

Heber-Katz E, Acha-Orbea H (1989) The V region hypothesis: evidence from autoimmune encephalomyelitis. Immunol Today 10:164–169

Howell MD, Winters ST, Olee T, Powell HC, Carlo DJ, Brostoff SW (1989) Vaccination against experimental allergic encephalomyelitis with T cell receptor peptides. Science 246:668–670

Kumar V, Sercarz EE (1993) The involvement of T cell receptor peptide-specific regulatory CD-4+ T cells in recovery from antigen-induced autoimmune disease. J Exp Med 178:909–916

Kumar V, Sercarz E (1998) Induction or protection from experimental autoimmune encephalomyelitis depends on the cytokine secretion profile of TCR peptide-specific regulatory CD4 T cells. J Immunol 161:6585–6591

Kumar V, Tabibiazar-Geysen HM, Sercarz EE (1995) Immunodominant framework region 3 peptide from TCR V beta 8.2 chain controls murine experimental autoimmune encephalomyelitis. J Immunol 154:1941–1950

Kumar V, Stellrecht K, Sercarz EE (1996) Inactivation of T cell receptor-peptide-specific CD4 regulatory T cells induces chronic experimental autoimmune encephalomyelitis (EAE). J Exp Med 184:1609–1617

Kumar V, Aziz F, Sercarz EE, Miller A (1997) Regulatory T cells specific for the same framework 3 region of the Vbeta8.2 chain are involved in the control of collagen II-induced arthritis and experimental autoimmune encephalomyelitis. J Exp Med 185:1725–1733

Lider O, Karin N, Shinitzky M, Cohen IR (1987) Therapeutic vaccination against adjuvant arthritis using autoimmune T cells treated with hydrostatic pressure. Proc Natl Acad Sci USA 84:4577–4580

Lider O, Reshef T, Beraud E, Ben-Nun A, Cohen IR (1988) Anti-idiotypic network induced by T cell vaccination against experimental autoimmune encephalomyelitis. Science 239:181–183

Mitchison NA (1998) Introduction. In: Bock GR, Goode JA (eds) Immunological tolerance. Novartis Foundation Symposium 215. Wiley, Chichester, pp 1–4

O'Garra A, Steinman L, Gijbels K (1997) CD4^{+} T-cell subsets in autoimmunity, Curr Opin Immunol 9:872–883

Saruhan-Direskeneli G, Weber F, Meinl E, Petter M, Diegerich G, Hinkkanen A, Epplen JT, Hohlfeld R, Wekerle H (1993) Human T cell autoimmunity against myelin basic protein: CD4+ cells recognizing epitopes of the T cell receptor beta chain from a myelin basic protein-specific T cell clone. Eur J Immunol 23:530–536

Vandenbark AA, Hashim G, Offner H (1989) Immunization with a synthetic T-cell receptor V-region peptide protects against experimental autoimmune encephalomyelitis. Nature 341:541–544

Vandenbark AA, Chou YK, Witham RH, Mass M, Buenafe A, Liefeld D, Kavanagh D, Cooper S, Hashim GA, Offner H, Bourdette DN (1996a) Treatment of multiple sclerosis with T-cell receptor peptides: results of a double-blind pilot trial. Nat Med 2:1109–1115

Vandenbark AA, Hashim GA, Offner H (1996b) T cell receptor peptides in treatment of autoimmune disease: rationale and potential. J Neurosci Res 43:391–402

Van Paris L, Perez VL, Abbas AK (1998) Mechanisms of peripheral T cell tolerance. In: Bock GR, Goode JA (eds) Immunological tolerance. Novartis Foundation Symposium 215. Wiley, Chichester, pp 5–13

Waisman A, Ruiz PJ, Hirschberg DL, Gelman A, Oksenberg JR, Brocke S, Mor F, Cohen IR, Steinman L (1996) Suppressive vaccination with DNA encoding a variable region gene of the T-cell receptor prevents autoimmune encephalomyelitis and activates Th2 immunity. Nat Med 2:899–905

Zhang J, Vandevyer C, Stinissen P, Raus J (1995) In vivo clonotypic regulation of human myelin basic protein-reactive T cells by T cell vaccination. J Immunol 155:5868–5877

Zipp F, Kerschensteiner M, Dornmair K, et al (1998) Diversity of the anti-T-cell receptor immune response and its implications for T-cell vaccination therapy of multiple sclerosis. Brain 121:1395–1407

12 Regulatory Aspects of Cancer Gene Therapy and DNA Vaccination

K. Cichutek

12.1 Introduction

The term gene therapy is used in Germany as a general term for human somatic gene and (genetically modified) cell therapy. Gene therapy encompasses the use of recombinant drugs in humans. These drugs are, on the one hand, naked DNA, viral, or non-viral vectors in vivo, or, on the other hand, genetically modified cells in vivo. The active ingredient is the gene product made in vivo, and thus, indirectly, also the gene to be delivered. The vector, if used, may be considered as a component of the drug. The formulation may contain additional ingredients.

Definition: Gene therapy and somatic cell therapy products used in vivo are medicinal products according to the German Drug Law (AMG; "Arzneimittelgesetz"). They include DNA, viral, or non-viral vectors, and genetically modified autologous, allogeneic, or xenogeneic cells (used in vivo) and are often summarized under the term "gene therapy drugs" in Germany. No official definition of the term "gene therapy drug" has been included in the AMG (up to the ninth supplementary version).

Generally, ready-prepared drugs are used in clinical trials of phase I–III to learn about their safety, efficacy, and potential environmental risks associated with their use. The data collected form the basis for an application for marketing authorization which, if granted, allows the standard use of the drug. In contrast, individually prepared drugs do not require marketing authorization in order to be used in humans. For example, autologous modified cells are generally only used in the homologous patient. Nevertheless, current German practice, as proposed in the GT guidelines of the German Medical Association and in the final report of the working group "Bund/Länder-Arbeitsgemeinschaft Somatische Gentherapie", is to also use individually prepared drugs under the provisions of § 40 AMG, i.e., in clinical trials.

Regulation of gene therapy drug use in humans prior to marketing authorization is mainly provided by the AMG and the professional law of physicians. Application of genetically modified organisms (GMOs) and therefore of gene therapy drugs in humans is not regulated by the German Gene Technology Law (GenTG). Approval of deliberate release according to the GenTG is not required. The regulations are identical for gene therapy drugs and other drugs.

12.2 Preclinical Research

Experimental work in gene therapy including the construction, use, storage, and inactivation of vectors, genetically modified bacterial or mammalian cells, or animals has to be conducted according to the GenTG ("Gentechnikgesetz"; transformation of Council Directives 90/219/EEC and 90/220/EEC). Basically, all experiments therefore are to be performed in gene technology laboratories under contained use. If only risk group 1 organisms are used, this only involves notification of the competent authority, which is different for each Land in Germany. This is true for the use of naked DNA or non-viral vectors and genetically modified cells assuming that genetic sequences void of any potential pathogenicity for humans or animals would be used. Work with risk group 2 organisms, such as adenoviral and retroviral (including lentiviral) vectors, has to be performed in safety level 2 laboratories which have to be registered and approved. Each line of experiment involving the use of GMOs also has to be approved. Similar preconditions are given for work with risk group 3 GMOs, which could for example include hybrid vectors derived from adeno- and lentiviruses.

Generally in Germany, experiments involving the use of GMOs have to be performed in laboratories or animal facilities of one of four safety levels (S1–S4), which are accordingly equipped. Laboratory approval is given by the competent authority of the Land ("Bundesland") for the GenTG. As experiments in safety level 1 laboratories only have to be documented and the competent authority has to be notified, the experiments can be started as soon as the authority has received the notification. As experiments falling under higher safety levels need additional approval by the same authority, experiments normally can be started about 3 months or less after application.

The Central Commission for Biological Safety (ZKBS; "Zentrale Kommission für die Biologische Sicherheit", secretariat located at the Robert-Koch-Institut, Berlin) provides a list containing the safety level classifications of "standard" vectors or plasmids and GMOs and is, in some cases (for example, approval of safety level 3 operations), to be consulted by the competent authority of the Land for the GenTG. Thus, advice about the approval of gene laboratories and genetic experiments is given by the competent authority of the Land where the contained use facility is located.

Other laws and regulations that may apply (which are executed by different competent authorities of the Land where the laboratory is located) include the law on epidemics ("Bundesseuchengesetz", to be replaced by the "Infektionsschutzgesetz"), the law on animal protection ("Tierschutzgesetz"), the law on human embryo protection ("Embryonenschutzgesetz"), the radiation protection ordinance ("Strahlenschutzverordnung"), and the ordinance on the use of hazardous substances ("Gefahrstoffverordnung").

12.3 Manufacture

Drugs for clinical use have to be produced according to Good Manufacturing Practice (GMP). GMP has to be implemented as defined in the "Operation Ordinance for Pharmaceutical Entrepreneurs (PharmBetrV; "Betriebsverordnung für pharmazeutische Unternehmer"). Also generally relevant are the GMP guidelines issued by the WHO, the European Community, and PIC. Good Laboratory Practice (GLP) implemented by the German law on the use of chemical substances (ChemG; "Chemikaliengesetz") is required in certain cases, for example, for analyses relevant to the safety of the drugs, according to the German law on the use of chemical substances. Manufacturing authorization is necessary for those manufacturers who intend to commercially or professionally distribute the drug (and/or the active ingredient) to others (Fig. 1). It is granted by the authority of the Land, where the facility is located, competent for the AMG (§ 13 AMG). This authority is also responsible for inspections and supervision (§ 64 AMG). Notification of the competent authority of the Land for the AMG is necessary beforehand for companies and establishments (also clinical departments) which develop, manufacture, test, package drugs, or subject them to clinical trials (§ 67 AMG).

Manufacturing authorization for drugs which are blood products (such as genetically modified blood stem cells) or vaccines (such as genetically modified tumor cells), is given under consultation with the Paul-Ehrlich-Institut. This involves a site visit from a member of the competent authority together with a member of the Paul-Ehrlich-Institut before authorization is granted. According to the drug law in its newer version (§ 14 AMG), the production and testing of the drug intended to be used in the clinic has to be performed according to the established standards of science and technology. Therefore, respective prescriptions for production and testing have to be included in the application and will be reviewed.

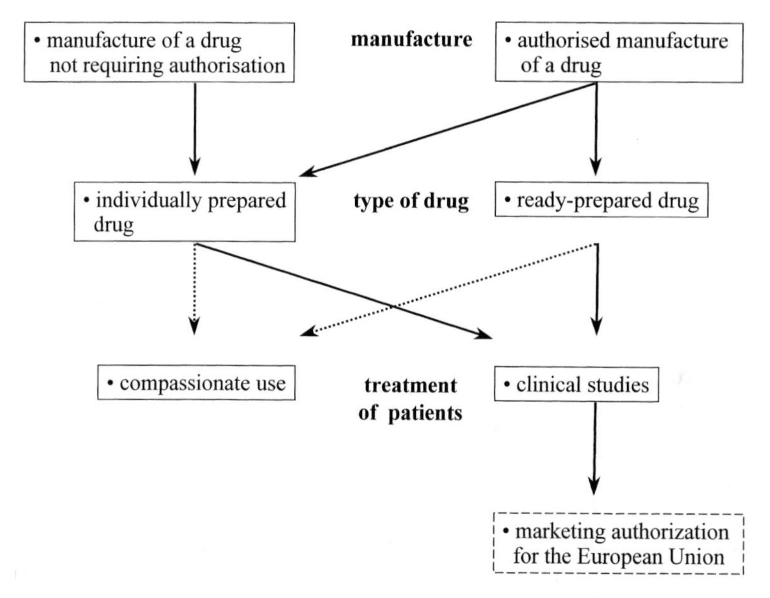

Fig. 1. Manufacture, clinical trial and marketing authorisation of gene therapy drug

12.4 Clinical Trial ("Klinische Prüfung")

In the guidelines "Richtlinien zum Gentransfer in menschliche Koerperzellen", published by the German Medical Association the requirement to use gene therapy drugs during clinical trials only is stressed (Fig. 1). Clinical trials can only be conducted, if certain requirements are met (see §§ 40, 41 AMG). Positive appraisal of a local, independent ethics committee formed according to the law of the Land, where the trial is performed, is required before initiation of a clinical trial (see § 40 (1) AMG for exception). If the trial is going to be performed in different clinics, all relevant ethics committees have to give their appraisal. No IND approval by a higher federal authority is necessary except for the submission of a complete set of certain documents (see below).

Notification of the competent authority of the Land for the AMG and deposition of the clinical study plan is required according to § 67 AMG. This authority is also responsible for inspections and supervision of the trial according to § 64 AMG. If necessary, samples of the material produced or tested can be taken.

Submission of documents (presentation according to § 40 AMG; forms available by Internet: http://www.dimdi.de/germ/amg/klifo.htm) including the positive appraisal of the local ethics committee(s) and the pharmacological/toxicological data (see German Directive "Arzneimittelprüfrichtlinien" for content) is to the competent federal higher authority for the AMG. The competent authority is either the Paul-Ehrlich-Institut, Langen, for gene therapy drugs which are vaccines or blood preparations, or the Federal Institute for Drugs and Medical Devices (BfArM), Berlin, for other gene therapy drugs (see § 77 AMG). The trial can only be initiated after written confirmation of the competent higher authority that all documents required have been received.

A positive appraisal is given by the Commission for Somatic Gene Therapy of the German Medical Association (KSG-BÄK, "Kommission für somatische Gentherapie", Koeln). This appraisal is required according to the professional law of physicians [see "Richtlinien zum Gentransfer in menschliche Koerperzellen", Deutsches Aerzteblatt 92, Heft 11, B-583-B588 (1995)]. The Commission is giving advice to the local ethics committees about issues related specifically to gene therapy. Members of the Commission are currently scientific experts in the fields of clinical gene therapy and so-called vectorology as well as an expert in ethics.

Registration of the clinical trial and patients involved with the German Gene Therapy Register (DGTR, "Deutsches Gentherapie-Register") is requested at the German Working Group for Gene Therapy (DAG-GT), and the publication of the protocol is recommended. The registration is important in order to keep records of the gene therapy trials going on.

12.5 Contacts and Further Information

The competent higher federal authorities involved in gene therapy can be contacted for further information. The author is a member of the Commission for Somatic Gene Therapy and was thus recruited from the Department of Medical Biotechnology of the Paul-Ehrlich-Institut to give advice on questions of drug safety and pharmacology

(Prof. Dr. K. Cichutek, Paul-Ehrlich-Str. 51–59, D-63225 Langen, Tel.: +49-6103-775307, Fax: +49-6103-771255, e-mail: cickl@pei.de). The Paul-Ehrlich-Institut also arranges informal discussion on the design of clinical gene therapy trials as well as questions of preclinical testing of such drugs and the biological monitoring with a view to drug licensing. The expert of the Federal Institute for Drugs and Medical Devices, Dr. U. Kleeberg, can also be contacted for further information (Seestr. 10–11, D-13353 Berlin, Tel.: +49-30-45483356, Fax: +49-30-45483332, e-mail: u.kleeberg@bfarm.de). However, the competent authorities of the Land are in charge of authorizing and supervising the drug manufacture and the clinical trials, and they can give advice on related questions.

The secretariat of the Commission for Somatic Gene Therapy (Wissenschaftlicher Beirat der Bundesaerztekammer, Herbert-Lewin-Str. 1, D-50931 Koeln, Tel.: +49-221-40040, Fax: +49-221-4404386, e-mail: dezernat6@baek.dgn.de) can be asked for advice on questions related to the application. The secretariat of the Commission can also arrange hearings during which general questions related to gene therapy can be discussed. For general information, the German Working Group for Gene Therapy (contact: Dr. M. Hallek, Muenchen, e-mail: hallek@lmb.uni-muenchen.de) is available.

12.6 Overviews on German Gene Therapy Regulations

Overviews on and introduction to German gene therapy regulations can be found in:

- Lindemann A, et al (1995) Guidelines for the design and implementation of clinical studies in somatic cell therapy and gene therapy. J Mol Med 73:207–211
- Abschlußbericht der Bund/Länder-AG "Somatische Gentherapie" (Final report of the federal ad hoc working group "Somatic Gene Therapy"). In: Eberbach/Lange/Ronellenfitsch (Hrsg.) Recht der Gentechnik und Biomedizin, GenTR/BioMedR, Teil II, F., Loseblattwerk, Heidelberg, Stand: 19. Erg.Lfrg., Dezember 1997
- Cichutek K, Krämer I (1997) Gene therapy in Germany and in Europe: regulatory issues. Qual Assur J 2:141–152

A brief summary of German gene therapy regulations is also available on the Euregenethy website:
(http://193.48.40.240/www/euregenethy/reg/Germanfront).

12.7 Guidelines for Cancer Gene Therapy and DNA Vaccination

Most clinical gene therapy trials in the European Union have been performed in the UK, France, and Germany. The system of regulations is similar in these member states except for the lack of an IND-like system in Germany. In addition to the appraisal of a local ethics committee, the appraisal of a more specialized central committee is required or at least optional. Documents have to be presented to competent authorities such as the Medical Control Agency (MCA) in the UK, the Agence du Medicament in France, and one of the two competent federal higher authorities in Germany. General information about the documents to be submitted is available, however, the format and, most likely, the contents of such documentation is probably different. Harmonization of drug regulations prior to licensing between member states could lead to mutual recognition of appraisals or authorization of clinical trials in European member states and the establishment of a more centralized one-door-one-key system for clinical trial initiation.

The main pharmacological/toxicological data to be presented prior to using a gene transfer drug in humans are described in European Notes for Guidance which are written and edited by drafting groups for the Biotechnology Working Party of the Committee for the Evaluation of Proprietary Medicinal Products (CPMP). One of the relevant guidelines for cancer gene therapy is the Note for Guidance ("Gene therapy products-...") which is currently under revision. For the formation of a drafting group for such a guideline, experts from drug authorities in European member states are normally recruited and cooperate on the formation of such guidelines. Therefore, a general consensus concerning the preclinical investigations to be made and documented in applications for clinical trial initiation has already been formed in Europe. It may be worth while to harmonize, in addition, legislative measures in different member states. Concomitantly, the centralized process to obtain marketing authorization may be paralleled by a similar centralized process of obtaining authorization for clinical trial initiation.

For prophylactic DNA vaccination, the WHO has issued a guideline providing advice for the development and preclinical testing. Main issues discussed are the quality and purity of the DNA as well as theoretical risks, such as autoimmune reactions, induction of tolerance, and chromosomal integration of the DNA with certain consequences. Further guidance will probably be part of the updated version of the European Note for Guidance ("Gene therapy products-...") or may be issued in the form of a separate European Note for Guidance for DNA vaccines only.

Subject Index

Ernst Schering Research Foundation Workshop

Editors: Günter Stock
Monika Lessl

Vol. 1 *(1991)*: Bioscience⇆Society – Workshop Report
Editors: D. J. Roy, B. E. Wynne, R. W. Old

Vol. 2 (1991): Round Table Discussion on Bioscience⇆Society
Editor: J. J. Cherfas

Vol. 3 (1991): Excitatory Amino Acids and Second Messenger Systems
Editors: V. I. Teichberg, L. Turski

Vol. 4 (1992): Spermatogenesis – Fertilization – Contraception
Editors: E. Nieschlag, U.-F. Habenicht

Vol. 5 (1992): Sex Steroids and the Cardiovascular System
Editors: P. Ramwell, G. Rubanyi, E. Schillinger

Vol. 6 (1993): Transgenic Animals as Model Systems for Human Diseases
Editors: E. F. Wagner, F. Theuring

Vol. 7 (1993): Basic Mechanisms Controlling Term and Preterm Birth
Editors: K. Chwalisz, R. E. Garfield

Vol. 8 (1994): Health Care 2010
Editors: C. Bezold, K. Knabner

Vol. 9 (1994): Sex Steroids and Bone
Editors: R. Ziegler, J. Pfeilschifter, M. Bräutigam

Vol. 10 (1994): Nongenotoxic Carcinogenesis
Editors: A. Cockburn, L. Smith

Vol. 11 (1994): Cell Culture in Pharmaceutical Research
Editors: N. E. Fusenig, H. Graf

Vol. 12 (1994): Interactions Between Adjuvants, Agrochemical and Target Organisms
Editors: P. J. Holloway, R. T. Rees, D. Stock

Vol. 13 (1994): Assessment of the Use of Single Cytochrome P450 Enzymes in Drug Research
Editors: M. R. Waterman, M. Hildebrand

Vol. 14 (1995): Apoptosis in Hormone-Dependent Cancers
Editors: M. Tenniswood, H. Michna

Vol. 15 (1995): Computer Aided Drug Design in Industrial Research
Editors: E. C. Herrmann, R. Franke

Vol. 16 (1995): Organ-Selective Actions of Steroid Hormones
Editors: D. T. Baird, G. Schütz, R. Krattenmacher

Vol. 17 (1996): Alzheimer's Disease
Editors: J.D. Turner, K. Beyreuther, F. Theuring

Vol. 18 (1997): The Endometrium as a Target for Contraception
Editors: H.M. Beier, M.J.K. Harper, K. Chwalisz

Vol. 19 (1997): EGF Receptor in Tumor Growth and Progression
Editors: R. B. Lichtner, R. N. Harkins

Vol. 20 (1997): Cellular Therapy
Editors: H. Wekerle, H. Graf, J.D. Turner

Vol. 21 (1997): Nitric Oxide, Cytochromes P 450,
and Sexual Steroid Hormones
Editors: J.R. Lancaster, J.F. Parkinson

Vol. 22 (1997): Impact of Molecular Biology
and New Technical Developments in Diagnostic Imaging
Editors: W. Semmler, M. Schwaiger

Vol. 23 (1998): Excitatory Amino Acids
Editors: P.H. Seeburg, I. Bresink, L. Turski

Vol. 24 (1998): Molecular Basis of Sex Hormone Receptor Function
Editors: H. Gronemeyer, U. Fuhrmann, K. Parczyk

Vol. 25 (1998): Novel Approaches to Treatment of Osteoporosis
Editors: R.G.G. Russell, T.M. Skerry, U. Kollenkirchen

Vol. 26 (1998): Recent Trends in Molecular Recognition
Editors: F. Diederich, H. Künzer

Vol. 27 (1998): Gene Therapy
Editors: R.E. Sobol, K.J. Scanlon, E. Nestaas, T. Strohmeyer

Vol. 28 (1999): Therapeutic Angiogenesis
Editors: J.A. Dormandy, W.P. Dole, G.M. Rubanyi

Vol. 29 (2000): Of Fish, Fly, Worm and Man
Editors: C. Nüsslein-Volhard, J. Krätzschmar

Vol. 30 (2000): Therapeutic Vaccination Therapy
Editors: P. Walden, W. Sterry, H. Hennekes

Vol. 31 (2000): Advances in Eicosanoid Research
Editors: C.N. Serhan, H.D. Perez

Supplement 1 (1994): Molecular and Cellular Endocrinology of the Testis
Editors: G. Verhoeven, U.-F. Habenicht

Supplement 2 (1997): Signal Transduction in Testicular Cells
Editors: V. Hansson, F. O. Levy, K. Taskén

Supplement 3 (1998): Testicular Function:
From Gene Expression to Genetic Manipulation
Editors: M. Stefanini, C. Boitani, M. Galdieri, R. Geremia, F. Palombi

Supplement 4 (2000): Hormone Replacement Therapy
and Osteoporosis
Editors: J. Kato, H. Minaguchi, Y. Nishino

Supplement 5 (1999): Interferon:
The Dawn of Recombinant Protein Drug
Editors: J. Lindenmann, W.D. Schleuning

Supplement 6 (2000): Testis, Epididymis and Technologies
in the Year 2000
Editors: B. Jégou, C. Pineau, J. Saez

This series will be available on request from
Ernst Schering Research Foundation, 13342 Berlin, Germany